Standard of Ministry of Water Resources of
the People's Republic of China

SL 618—2013
Replace DL 5020—93

Code for Preparing Feasibility Study Report of Water and Hydropower Projects

Drafted by:
Water Resources and Hydropower Planning and Design General Institute, Ministry of Water Resources

Translated by:
China Water Resources Beifang Investigation Design and Research Co. Ltd.

China Water & Power Press
Beijing 2018

·北京·

图书在版编目（CIP）数据

水利水电工程可行性研究报告编制规程：SL618-2013 = Code for Preparing Feasibility Study Report of Water and Hydropower Projects SL618-2013：英文/中华人民共和国水利部发布. -- 北京：中国水利水电出版社，2018.4
ISBN 978-7-5170-6456-5

Ⅰ.①水… Ⅱ.①中… Ⅲ.①水利水电工程－可行性研究－研究报告－编制－规程－中国－英文 Ⅳ.①TV-65

中国版本图书馆CIP数据核字(2018)第105318号

书　　名	Code for Preparing Feasibility Study Report of Water and Hydropower Projects SL 618—2013
作　　者	中华人民共和国水利部　发布
出版发行	中国水利水电出版社 （北京市海淀区玉渊潭南路1号D座　100038） 网址：www.waterpub.com.cn E-mail：sales@waterpub.com.cn 电话：（010）68367658（营销中心）
经　　售	北京科水图书销售中心（零售） 电话：（010）88383994、63202643、68545874 全国各地新华书店和相关出版物销售网点
排　　版	中国水利水电出版社微机排版中心
印　　刷	北京瑞斯通印务发展有限公司
规　　格	140mm×203mm　32开本　4.375印张　152千字
版　　次	2018年4月第1版　2018年4月第1次印刷
定　　价	**320.00元**

凡购买我社图书，如有缺页、倒页、脱页的，本社营销中心负责调换

版权所有·侵权必究

Introduction to English Version

Department of International Cooperation, Science and Technology of Ministry of Water Resources, P. R. China (hereinafter DICST) has the mandate of managing the formulation and revision of water technology standards in China.

Translation of this standard from Chinese into English is organized by DICST in accordance with due procedures and regulations applicable in China.

The English version of this standard is identical to its Chinese original SL 618—2013, which was formulated and revised under the auspices of DICST.

Translation of this standard is undertaken by China Water Resources Beifang Investigation Design & Research Co. Ltd.

Translation task force includes DU Leigong, SHAO Jiannan, YANG Haiyan, CHEN Shaosong, HAO Fuliang, XU Xian'e, DONG Keqing, MA Rentao, LI Liwei, SUN Fuhang, ZHAO Limin, CHENG Jing, MA Shilong, WU Guilan, LYU Chuanliang, YUAN Xuean, LIU Zhansheng, SUN Tianqing, WANG Hao, WENG Jianping, CHI Shouxu, LU Yonghua, ZHAO Dongliang, WANG Wei, ZHANG Weibo, FENG Shaohui, HU Nan, YU Yunfei, DING Xiuxia, and LI Miao.

This standard is reviewed by LI Guangcheng, ZHANG Zhi, MENG Zhimin, JIN Hai, LIU Yongfeng, XU Zeping, GU Liya, LIN Ning, and SUN Feng.

Department of International Cooperation, Science and Technology
Ministry of Water Resources, P. R. China

Foreword

According to the Development and Revision Plan of technical standards for water resources issued by the Ministry of Water Resources, in light of the requirements of SL 1—2002 *Specification for the Drafting of Technical Standards of Water Resources*, revision is made to DL 5020—93 *Specifications on Compiling Feasibility Study Report of Water Conservancy and Hydropower Projects* jointly issued by the former Ministry of Power Industry and Ministry of Water Resources.

This standard comprises 17 chapters and 2 annexes with the main technical contents as follows: general provisions, executive summary, hydrology, engineering geology, purposes and scale of project, project layout and structures, electromechanical equipment and hydraulic steel works, construction planning, land acquisition and resettlement, environmental impact assessment, water and soil conservation, labor safety and industrial hygiene, energy saving evaluation, project management, cost estimate, economic evaluation, social stability risk analysis, etc.

The main revisions are as follows:

—Adjusting the working requirements, and emphasizing schemes comparison and selection in different disciplines affecting the project scale and cost.
—Adding the contents and requirements on structures for levee and river training works, irrigation and drainage, water supply, navigation, etc.
—Upgrading the requirements on land acquisition, inventory investigation and resettlement plan.

—Supplementing the contents of water and soil conservation, labor safety and industrial hygiene, energy saving evaluation, etc.
—Putting forward specific requirements on project management according to the project category.
—Adding the calculation of loan capacity, and schemes comparison and selection in the financial evaluation.
—Adding the content and requirements of social stability risk analysis.

This standard replaces the previous versions
—SL 123—84
—DL 5020—93

This standard is approved by Ministry of Water Resources of the People's Republic of China.

This standard is interpreted by Department of Programming and Planning, Ministry of Water Resources.

This standard is chiefly drafted by Water Resources and Hydropower Planning and Design General Institute, Ministry of Water Resources.

This standard is jointly drafted by Jianghe Consulting Center for Water Resources and Hydropower.

This standard is published and distributed by China Water & Power Press.

Chief drafters of this standard are WANG Hong, WANG Zhiming, DUAN Hongdong, ZENG Zhaojing, DONG Anjian, MEI Mianshan, CHEN Wei, WANG An'nan, TIAN Kejun, LIN Zhao, CHEN Deji, LI Xiaoyan, SHAO Jiannan, SHI Haifeng, WU Yunping, CHEN Jianjun, SUN Shuangyuan, SI Fu'an, LI Xianshe, JIANG Xiao, JIANG Jiaquan, YANG Qing, ZHU Chuanbao, ZHU Dangsheng, LIU Jiang, NIU

Wusheng, XU Ji, HE Ding'en, PAN Shangxing, ZHOU Yimei, LEI Xingshun, HU Yuqiang, YIN Xunfei, WANG Zhiguo, BU Shuhe, CHEN Shuwen, YUN Qinglong, WANG Yingren, LI Weitao, LI Guangcheng, JU Zhanbin, WANG Pengji, GUAN Zhicheng, REN Tiejun, LIU Congning, ZHAO Xuemin, LIU Zhihong, MA Yugan, Chen Qinglian, HAN Zengfen, LIU Yixin, LIU Shuqiu, XU Gonglu, SHEN Demin, and LIU Hairui.

The leading experts of the technical review meeting of this standard are ZHU Erming and LIU Zhiming.

The format examiner of this standard is CHEN Hao.

Contents

Introduction to English Version
Foreword
1 General Provisions 1
2 Executive Summary 5
3 Hydrology 9
   3.1 River Basin Overview 9
   3.2 Meteorology 9
   3.3 Basic Hydrologic Data 9
   3.4 Runoff 10
   3.5 Flood 11
   3.6 Drainage (Waterlogging Evacuation) Modulus and
        Discharge 12
   3.7 Sediment 13
   3.8 Rating Curve 13
   3.9 River Water Level and Tidal Level 13
   3.10 Surface Evaporation and Ice Regime 14
   3.11 Automatic Hydrologic Data Acquisition and Transmission
         System 15
   3.12 Figures and Tables 15
4 Engineering Geology 17
   4.1 General Description 17
   4.2 Regional Tectonics and Ground Motion Parameters 17
   4.3 Engineering Geology of Reservoir Area 17
   4.4 Engineering Geology of Dam (Sluice) Site 18
   4.5 Engineering Geology of Water Releasing Structure Site 19
   4.6 Engineering Geology along Diversion Structure

	for Power Generation	19
4.7	Engineering Geology of Powerhouse and Pump Station Site	20
4.8	Engineering Geology of Navigation and Fish Pass Structures Site	20
4.9	Engineering Geology of River Diversion and Closure Works Site	20
4.10	Engineering Geology along Water Conveyance Routes	21
4.11	Engineering Geology of Levee and River Training Works	21
4.12	Hydrogeology of Irrigation Area	22
4.13	Natural Construction Materials	22
4.14	Figures and Tables	23
5	Purposes and Scale of Project	24
5.1	Necessities and Purposes of Project Development	24
5.2	Flood Control	27
5.3	Waterlogging Control	30
5.4	River Channel and Estuary Training	32
5.5	Irrigation	33
5.6	Water Supply	36
5.7	Hydropower	39
5.8	Multi-Purpose Project	41
5.9	Reinforcement, Renovation and Extension Projects	47
5.10	Figures, Tables and Appendices	48
6	Project Layout and Structures	50
6.1	Design Basis	50
6.2	Project Grade and Standard	50
6.3	Selection of Project Site and Route	50
6.4	Selection of Structural Types	51
6.5	General Project Layout	52
6.6	Water Retaining Structures	52

6.7	Water Releasing Structures	53
6.8	Diversion Structures for Power Generation	54
6.9	Powerhouse (Pump House) and Switchyard (Substation)	55
6.10	Slope Works	55
6.11	Irrigation and Drainage Structures	56
6.12	Water Supply Works	57
6.13	Navigation and Fish Passing Structures	58
6.14	Levee and River Training Works	59
6.15	Access Works	60
6.16	Engineering Safety Monitoring	61
6.17	Reinforcement of Existing Projects	61
6.18	Figures, Tables and Appendices	62
7	Electromechanical Equipment and Hydraulic Steel Works	64
7.1	Hydraulic Machinery	64
7.2	Electrical System	65
7.3	Hydraulic Steel Works	66
7.4	Heating, Ventilation and Air Conditioning (HVAC)	66
7.5	Fire Fighting	67
7.6	Figures and Tables	67
8	Construction Planning	69
8.1	Construction Conditions	69
8.2	Selection and Exploitation of Borrow Area/Quarry	69
8.3	River Diversion and Closure During Construction	70
8.4	Construction of Main Works	70
8.5	Construction Transportation and General Construction Layout	71
8.6	General Construction Schedule	72
8.7	Figures, Tables and Appendices	72

9	Land Acquisition and Resettlement	74
9.1	General Description	74
9.2	Scope of Land Acquisition	74
9.3	Land Acquisition Objects	74
9.4	Rural Resettlement	75
9.5	Town (Fair) Relocation	77
9.6	Disposition of Industrial Enterprises	77
9.7	Disposition of Specific Works	77
9.8	Protection Works	78
9.9	Clearance of Reservoir Basin	78
9.10	Figures, Tables and Appendices	78
10	Environmental Impact Assessment	80
10.1	General Description	80
10.2	Investigation and Assessment of Environmental Status	80
10.3	Environmental Impact Prediction and Assessment	80
10.4	Environmental Protection Countermeasures	82
10.5	Environmental Management and Monitoring	83
10.6	Comprehensive Assessment Conclusion	83
10.7	Figures and Appendices	83
11	Water and Soil Conservation	84
11.1	General Description	84
11.2	Water and Soil Conservation Assessment for Main Works	84
11.3	Responsible Area and Zoning of Soil Erosion Control	84
11.4	Prediction of Soil Erosion	85
11.5	Standard and General Layout of Soil Erosion Control	85
11.6	Design of Zoning Control Measures	85
11.7	Water and Soil Conservation Management and Monitoring	86
11.8	Figures and Appendices	86
12	Labor safety and Industrial Hygiene	87

12.1	Analysis of Dangerous and Harmful Factors	87
12.2	Labor Safety Measures	87
12.3	Industrial Hygiene Measures	88
12.4	Safety and Health Assessment	88

13 Energy Saving Evaluation — 89

13.1	Design Basis	89
13.2	Energy Consumption Analysis	89
13.3	Energy Conservation Measures	89
13.4	Energy Conservation Effect Evaluation	90

14 Project Management — 91

14.1	Project Management System	91
14.2	Project Operation Management	91
14.3	Scope of Management and Protection	92
14.4	Management Facilities and Equipment	92

15 Cost Estimate — 93

15.1	General Description	93
15.2	Principle and Contents of Preparation	93
15.3	Results of Cost Estimate	95
15.4	Cost Comparison and Analysis	99

16 Economic Evaluation — 100

16.1	General Description	100
16.2	Expenditure Estimation	100
16.3	National Economic Evaluation	100
16.4	Fund Financing Plan	101
16.5	Financial Evaluation	102
16.6	Figures and Tables	102

17 Risk Analysis of Social Stability — 104

17.1	Basis of Preparation	104
17.2	Risk Identification	104
17.3	Risk Factors Analysis	104

17.4	Risk Prevention and Mitigation Measures	105
17.5	Risk Analysis Conclusions	105
Annex A	Format of Feasibility Study Report	106
Annex B	Format of the Table of Project Characteristics at Feasibility Study Stage	108
Explanation of Wording		127

1 General Provisions

1.0.1 This code is formulated to standardize the principles, contents, and levels of the design in preparing the feasibility study reports for water and hydropower projects.

1.0.2 This code is applicable to preparing feasibility study reports for construction, renovation, and extension of large and medium-sized water and hydropower projects. Some contents may be omitted depending on the characteristics of the project. This code may also serve as a reference for preparing the feasibility study reports on reinforcement projects.

1.0.3 The feasibility study report shall be prepared based on the approved project proposal. For projects without performing project proposal study, the feasibility study reports shall be prepared based on the approved river basin (reach) development plan, regional comprehensive development plan or professional plan, and specific plan.

1.0.4 In preparation of the feasibility study report, national regulations shall be followed. The construction conditions of the project shall be investigated and surveyed in accordance with relevant technical standards. Comparison of alternatives shall be conducted based on reliable data, and comprehensive justification shall be made in view of technique, economy, society, environment, water and energy conservation, etc. to evaluate the feasibility of the project. Emphasis shall be placed on the project scale, technical scheme, land acquisition, resettlement, environment, cost, and economic evaluation. Specific studies shall be made on vital technical issues.

1.0.5 Contents and levels of the feasibility study report shall meet the following requirements:

1 Justify the project necessities, and determine the purposes and their prioritization for a multipurpose project.

2 Determine major hydrological parameters and results.

3 Ascertain main engineering geological conditions affecting schemes comparison and selection, basically ascertain engineering geological conditions of main structures site, and evaluate the engineering geological problems. Conduct a detailed investigation into natural construction materials.

4 Determine scale and general layout of the project.

5 Select project sites (dam site, sluice site, powerhouse site, pump station site, routes etc.)

6 Determine the project grade and design criteria, select basic type of dam, and basically select general layout of the project and types of other main structures.

7 Basically select the types and arrangement of electromechanical equipment, hydraulic steel works, and other major auxiliary equipment.

8 Preliminarily determine the design scheme and main facilities of fire-fighting system.

9 Select access and transportation scheme, borrow area and quarries, mode of river diversion, and layout of diversion structures. Basically select construction methods of main works and general layout for construction. Propose critical milestones and proposal for implementation in stages and basically determine the total construction duration.

10 Determine the scope of land acquisition, identify objects to be inundated, basically determine the resettlement plan and estimate the cost of compensation for the land acquisition and resettlement.

11 Carry out the prediction and assessment on environmental impacts for main environmental elements, determine the environmental protection countermeasures, and estimate the cost of environmental protection.

12 Carry out the assessment on water and soil conservation for main works design, determine the obligation scope of the soil erosion control, measures, and monitoring scheme of water and soil conservation, and estimate the cost of the water and soil conservation.

13 Preliminarily determine the design scheme of labor safety and industrial hygiene, and basically determine main measures.

14 Define types and quantity of energy consumption, indices of energy consumption, and design principle. Basically determine energy saving measures.

15 Determine the type and nature of management agency, institutional settings, scope of management, protection, etc.

16 Prepare the cost estimate.

17 Analyze the project benefit, cost, and debt service capability, propose the financing program, analyze main economic indices, and evaluate the economic rationality and financial feasibility of the project.

1.0.6 The feasibility study report shall be prepared in such frame of *Executive Summary* as Chapter 1, the following chapters shall be prepared according to the requirements of Chapter 3 to 17 herein in turn, and *Conclusions and Suggestions* as the last chapter. Words in the report shall be standard and accurate, the contents shall be concise and the drawings shall be clear and complete. See Annex A for the format of the feasibility study report of water and hydropower projects.

1.0.7 The feasibility study report of water and hydropower projects may include the following Appendixes:

1 Approved documents for project proposal report and other relevant important documents.

2 Special subject justification, review meeting minutes and comments.

3 Hydrological analysis report.

4 Engineering geologic investigation report.

5 Special report on project scale justification.

6 Report on compensation for land acquisition and resettlement plan.

7 Environmental impact assessment report (form).

8 Report on water and soil conservation plan.

9 Special report on debt service capacity calculation.

10 Special reports on other key technologies.

1.0.8 Provisions stipulated not only in this code but also in the prevailing national standards shall be complied with.

2 Executive Summary

2.0.1 In the Preface, the following contents shall be briefed:

1 Situation in connection with project location, socio-economic status, and conditions of nature, geography and resources.

2 National socio-economic development plan in the region where the project is located, and the role and function of the project in river training and development, and in the river master plan.

3 General situation of previous works, the main conclusions, the review and approval comments on the project proposal report.

4 Necessity and urgency of the project.

2.0.2 For hydrology, brief the overview of hydrology and meteorology, basic hydrological data and main hydrological calculation results in the river basin (reach) and region where the project is located.

2.0.3 For engineering geology, brief the overview of regional geology, the investigation results of hydrogeology, engineering geology and natural construction materials, the evaluation and conclusions of major geological problems, the main geotechnical parameters, and the engineering treatment suggestions.

2.0.4 For project purposes and scale, the following items shall be briefed:

1 Project purposes and their prioritization, overall utilization benefits and main technical-economic indices.

2 Scope of works, general layout, components to be constructed, and their scales.

2.0.5 For project layout and structures, the following items shall be briefed:

1 Project grade and standard.

2 Project site (route), model selection of main structures, general layout of project, layout and type of main structures, foundation treatment measures, etc.

2.0.6 For electromechanical equipment and hydraulic steel works, the following items shall be briefed:

1 Model, quantity, main parameters, and arrangement of major electromechanical equipment and hydraulic steel works.

2 Connection to power system, main electrical connection, supervisory control and data acquisition (SCADA) system, and communication system.

2.0.7 For construction planning, the following items shall be briefed:

1 Construction conditions and materials.

2 Standards and schemes of river diversion and closure.

3 Construction methods of main works and general construction layout.

4 Critical milestones and total construction duration.

2.0.8 For land acquisition and resettlement, the following items shall be briefed:

1 Scope, principle, and criterion of land acquisition.

2 Methods, contents, and results of main inventory investigations.

3 Rural resettlement plan, town (fair) relocation and construction plan, the treatment plan for industry and mining enterprises and specialized projects, and the plan for restoration and protection works.

2.0.9 For environmental impact assessment, the following items shall be briefed:

1 Conclusions of the environmental investigation and assessment, and the main protection goals in the region where the project is located.

2 Prediction and assessment conclusions of environmental impact and countermeasures.

2.0.10 For water and soil conservation, the following items shall be briefed:

1 Conclusions, requirements and suggestions of the water and soil conservation assessment for main works.

2 Obligation scope and prediction result of soil erosion control works.

3 Standard, target, zoning, general layout, and main quantities of soil erosion control works.

2.0.11 For labor safety and industrial hygiene, brief the standards and protective schemes for main structures, production equipment, working environment, monitoring device, and safety facilities related to the labor safety and industrial hygiene.

2.0.12 For energy saving evaluation, brief the basis and principle of the design, measures of energy saving, and the conclusions of energy consumption analysis and comprehensive evaluation.

2.0.13 For project management, brief the property of management agency, organization structuring scheme, personal staffing, scope of management and protection, main management facilities, management expenses and their sources, etc.

2.0.14 For cost estimate, brief the principle, basis and results for cost estimate of the main works, land acquisition and resettlement, environmental protection, water and soil conservation, and brief the change in cost and causes compared with that of previous stage.

2.0.15 For economic evaluation, brief the benefits estimation, national economic evaluation, financing scheme, and main methods and conclusions of financial evaluation.

2.0.16 Brief the conclusions on risks analysis of social stability.

2.0.17 For conclusions and suggestions, brief the main conclusions and suggestions of feasibility study of the project, and describe the main difference of the main conclusions from the review and approval comments at the project proposal stage.

2.0.18 Documents related to the project construction should be attached to this chapter.

2.0.19 The following figures and tables should be attached:

1 Salient features of the project (see template in Annex B).

2 Sketch of geographic location.

3 Sketch of development status and plan of the river basin (reach) where the project is located.

4 Sketch of general layout of the project.

5 Engineering geological drawings of main structures.

6 General layout of the project.

7 Layout and profiles of major structures.

8 Geographic map of connection to power system.

9 General layout of construction.

10 Sketch map of reservoir inundation.

3 Hydrology

3.1 River Basin Overview

3.1.1 Describe the natural geography, the characteristics of river basin and river, water and soil conservation of the basin where the project is located.

3.1.2 Describe the location and main tasks of water and hydropower projects constructed and under construction on the river where the project is located.

3.2 Meteorology

3.2.1 Describe distribution and observation situations of the meteorological gauges and stations, in the river basin and neighboring areas of the project.

3.2.2 Describe characteristics of meteorological elements in the river basin and the area where the project is located.

3.3 Basic Hydrologic Data

3.3.1 Describe distribution of the hydrological stations in the river basin, the basin characteristics of the project site, design basis stations and reference stations.

3.3.2 Describe the observation items and period, control characteristics of river reach and elevation system of the design basis stations and reference stations, the measuring methods and precision of water level, stream flow and sediment, the processing of hydrological data, etc.

3.3.3 Describe the main problems in hydrometry and data compilation, changes after data recheck, and evaluate the reliability of basic hydrologic data.

3.4 Runoff

3.4.1 Brief the replenishment and sources of runoff, and the impact of upstream water and hydropower projects constructed and under construction on the runoff.

3.4.2 Make correction calculation for the runoff affected by human activities.

3.4.3 The runoff may be extrapolated and interpolated for the data with short record or discontinued record in some years.

3.4.4 Determine the calculation time interval (month, ten-day, and day) for runoff series based on the project characteristics and design requirements.

3.4.5 Analyze the consistency and representativeness of natural runoff series at hydrological stations, determine the natural runoff series of project site and interval, and propose annual monthly average stream flow and runoff distribution within a year.

3.4.6 Propose the annual runoff calculation results, and determine the calculation results after rationality analysis.

3.4.7 Analyze the characteristics of low-flow runoff, and carry out calculation on the low-flow runoff (period, month, and day), if necessary.

3.4.8 Analyze and determine the annual runoff for wet year, normal year and dry year if necessary.

3.4.9 For irrigation or water-supply project, describe the groundwater resource and its replenishment way, water table fluctuation, water quality, spatial distribution, available yield, etc. in related areas.

3.5 Flood

3.5.1 Brief the characteristics and causes of rainstorm, common rainstorm centers in the basin, observed and investigated magnitude and return period of heavy rainstorms.

3.5.2 Brief the cause, characteristics, and the spatial and temporal distribution of flood.

3.5.3 Describe the investigation and review results of historical flood, analyze and determine flood peak discharge, volume of the flood for a certain period, and return periods of historical flood and observed extraordinary flood.

3.5.4 Brief the influence of upstream existing water and hydropower projects on flood, and describe the work carried out on reversion, interpolation and extrapolation for flood series.

3.5.5 Describe the frequency flood calculation at project site as follows in light of available data:

1 When calculating the frequency flood according to streamflow series, describe the statistical principles for flood peak and volume series, number of years of flood series, the type of frequency curve and empirical frequency formula adopted; carry out flood frequency calculation; analyze and verify the rationality of calculation results; determine flood parameters and results at the project site, relevant cross – sections and intervals. Choose the typical flood event, and derive the design flood hydrograph.

2 When reckoning the frequency flood according to rainstorm data, describe the design rainstorm and rainfall – runoff model, verify the rationality of calculation results, and determine the frequency flood results.

3.5.6 Describe the calculation methods of reservoir inflow frequency flood and determine the results.

3.5.7 Describe the calculation methods of probable maximum storm and probable maximum flood, and determine the results of probable maximum flood subject to comprehensive analysis.

3.5.8 Describe the cause of phased frequency flood in flood season, propose the distribution diagram of monthly maximum streamflow within a year, describe the phases dividing and statistical principle of flood series, calculate the phased frequency flood, carry out rationality analysis and basically determine the calculation results of phased frequency flood. The results of yearly maximum frequency flood shall be adopted for the main flood season, and the phased frequency flood results may be used for the corresponding periods.

3.5.9 As required for preparing construction planning, describe the phases in the period of non-flood season and calculation methods of phased frequency flood, and basically determine the results of phased frequency flood.

3.5.10 Describe the position of flood control section adopted in determining the composition of flood area, and analyze the areal composition law of large flood and the condition of flood encountering. Describe methods of frequency flood area upstream the flood control section and typical flood years, magnifying method, flood routing parameters adopted. Provide the calculation results and carry out the rationality analysis, and determine the calculation results of frequency flood area composition.

3.5.11 Basically determine results of the flood of gullies or streams which are to be crossed with the water conveyance line and the frequency flood for crossing structures.

3.6 Drainage (Waterlogging Evacuation) Modulus and Discharge

3.6.1 Specify the river basin characteristics of drainage area, flood data or rainstorm data, calculation formula and parameters selection.

3.6.2 Provide calculation results and carry out the rationality analysis, design the modulus and discharge of drainage (waterlogging evacuation).

3.7 Sediment

3.7.1 Specify the sediment source and the influence of sediment retaining of upstream water and hydropower projects, and observed sediment series, and calculate the average annual sediment runoff of suspended load and bed load.

3.7.2 For projects with serious sediment problem, propose the particle size distribution curve, mean or median grain size and mineral composition of the suspended load.

3.8 Rating Curve

3.8.1 Specify location of the section, observed and investigated hydrologic data which are used to calculate the rating curve.

3.8.2 Specify methods for calculating the rating curve and extrapolating of high water level, and basically determine the rating curve of the section.

3.9 River Water Level and Tidal Level

3.9.1 When calculating the design river water level based on the design flow and using the rating curve, describe the methods and results of the design discharge calculation, the change of aggradation and degradation at the cross section and adopted rating curve, analyze and determine the design water level at the design section.

3.9.2 When the design level of rivers is analyzed and calculated directly based on observed water level data, describe the observed water levels and interpolation of the observed data, the length of water level series, evaluate the reliability, consistency and representativeness of the water level data series; provide the calculation results of water level frequency, analyze the rationality of calculation results, and determine the design level of the design section.

3.9.3 Specify the law of tide and its characteristic levels in the project area, analyze the cases of simultaneous occurrence of tide and flood, evaluate the reliability of the tidal level data (yearly maximum tidal level, yearly minimum tidal level, annual average tidal level) and the consistency and representativeness of the series, and determine the design tidal level and tidal level hydrograph. For sea-crossing projects, describe the law of sea current, and the sea depth along the project, determine the sea current direction and velocity distribution.

3.10 Surface Evaporation and Ice Regime

3.10.1 Describe the type of evaporation pan and the situation of observation, evaporation conversion coefficient for various evaporation pans, conversion coefficient of evaporation capacity for large water body and evaporation pan, and the series of observed and interpolated evaporations, and basically determine the average yearly, monthly evaporation of large water surface.

3.10.2 For the river reach with ice run hazard, describe the characteristics of ice regime, analysis and calculation of the ice regime for the project, analyze the potential ice problem during construction and operation of the project, and propose suggestions on measures of ice prevention and ice releasing.

3.11 Automatic Hydrologic Data Acquisition and Transmission System

3.11.1 Brief the hydrological and meteorological status and conditions of station network and sites, the observed data, locations of existing telemetry stations and the mode of communication networking in the design river basin.

3.11.2 Propose hydrological forecast scheme, scope of telemetry station network, site numbers, communication method, networking scheme, civil works, and equipment for all kinds of sites, estimate the investment.

3.12 Figures and Tables

3.12.1 The following figures may be attached to this chapter:

1 Map of watershed river systems (indicating locations of hydrological and meteorological stations, the project, the constructed and under-construction water and hydropower projects of large and medium-size scale).

2 Relevant interpolation and extrapolation figures of the runoff, flood, rainstorm and sediment.

3 Frequency curves on the annual (period) runoff and annual (period) precipitation.

4 Frequency curves on flood peaks, flood volumes or rainstorms.

5 Hydrographs of typical flood and design flood.

6 Particle size distribution curve of suspended load.

7 Chart of rating curve at main hydrologic stations and design cross-sections.

8 Other figures.

3.12.2 The following tables may be attached to this chapter:

1 Yearly statistical data on hydrometry at design basis stations.

2 Series (design basis stations, project site and interval) for yearly and monthly runoff (rainfall).

3 Series (design basis stations, project site and interval) for flood peak and volume (storm rainfall).

4 Hydrographs on typical flood and design flood.

5 Table of suspended sediment runoff.

6 Other tables.

4 Engineering Geology

4.1 General Description

4.1.1 Describe processes, major results, findings and conclusions, as well as review and evaluation comments on the geological and geotechnical investigation at previous stages.

4.1.2 Describe the contents of geological and geotechnical investigation and main workload completed at the current stage.

4.2 Regional Tectonics and Ground Motion Parameters

4.2.1 Brief regional geology of the project.

4.2.2 Evaluate regional tectonic stability and propose ground motion parameters.

4.3 Engineering Geology of Reservoir Area

4.3.1 Brief engineering geology in the reservoir area.

4.3.2 For reservoirs with potential leakage problem, identify the boundary conditions and the type of the leakage, estimate the amount of leakage, and propose the scope of leakage – control works.

4.3.3 For reservoirs with potential immersion problem, identify the critical buried depth of the groundwater, and predict the scope and degree of immersion and potential impacts.

4.3.4 Appraise the scope and volume of potential collapse, landslide and bank slump in the reservoir area, evaluate their stability and impacts, and propose monitoring suggestions.

4.3.5 Basically determine the main physical and mechanical parameters for rock mass and soil of the protective works in the reservoir area, and evaluate major engineering geological problems and environmental hydro-geological problems.

4.3.6 Predict the seismic triggering location and magnitude of the reservoir-induced earthquakes and evaluate their impact on the project. Propose the design scheme of seismic monitoring station/network if necessary.

4.3.7 Assess the environmental geological issues impacting on reservoir construction.

4.4 Engineering Geology of Dam (Sluice) Site

4.4.1 Brief engineering geology of alternative dam (sluice) sites.

4.4.2 Evaluate main engineering geological problems of alternative (sluice) sites and provide opinions on schemes selection of dam (sluice) sites from engineering geological point of view.

4.4.3 Basically determine main physical and mechanical parameters for rock mass and soil and evaluate the major engineering geological problems of the main project structures at the recommended dam (sluice) site.

4.4.4 For the recommended dam (sluice) site, evaluate the engineering geological suitability of all potential dam types, recommend preferred dam type from engineering geological point of view. Evaluate the major engineering geological problems of the recommended dam type, propose the competent dam foundation criteria, scope of seepage control treatments, and preliminary classification of foundation rock mass quality.

4.4.5 For the dam (sluice) to be constructed on soft foundation, apart from stipulations given in Article 4.4.4, it is also necessary to evaluate the slope stability of foundation pit and impacts of groundwater on the construction, and propose the bearing capacity of foundation and other relevant geological parameters for design of foundation reinforcement measures.

4.5 Engineering Geology of Water Releasing Structure Site

4.5.1 Brief engineering geology of water releasing structure site.

4.5.2 Evaluate the major engineering geological problems of alternative schemes. Provide engineering geological opinions on schemes selection at this stage.

4.5.3 Verify main physical and mechanical parameters for rock mass and soil. Evaluate stability of the weir foundation and excavation slopes and the expected engineering geological problems at the downstream energy dissipation area for the recommended spillway. Make preliminary engineering geological classification of the surrounding rock.

4.6 Engineering Geology along Diversion Structure for Power Generation

4.6.1 Brief engineering geology along diversion structure.

4.6.2 Evaluate the major engineering geological problems of alternative schemes and provide engineering geological opinions on schemes selection.

4.6.3 Basically determine main physical and mechanical parameters for rock mass and soil. Evaluate major engineering geological problems of foundation, slopes at inlet and outlet sections, and surrounding rock of tunnel for the recommended diversion structure. Make preliminary engineering geological classification of the surrounding rock.

4.7 Engineering Geology of Powerhouse and Pump Station Site

4.7.1 Brief engineering geology of powerhouse and pump station site.

4.7.2 Evaluate the major engineering geological problems of alternative schemes and provide engineering geological opinions on sites selection of powerhouse (pump station).

4.7.3 Basically determine main physical and mechanical parameters for rock mass and soil. Evaluate the major engineering geological problems of the sites for the recommended powerhouse and pump station.

4.7.4 For underground powerhouse, make preliminary engineering geological classification of tunnel/cavern surrounding rock. Evaluate the stability of surrounding rock and major engineering geological problems.

4.8 Engineering Geology of Navigation and Fish Pass Structures Site

4.8.1 Brief engineering geology of navigation and fish pass structures site.

4.8.2 Evaluate the major engineering geological problems of alternative schemes. Provide engineering geological opinions on schemes selection.

4.8.3 Basically determine main physical and mechanical parameters for rock mass and soil. Evaluate the major engineering geological problems of the recommended scheme.

4.9 Engineering Geology of River Diversion and Closure Works Site

4.9.1 Brief engineering geology of site for river diversion and closure works and cofferdam.

4.9.2 Evaluate sliding stability, leakage, and permeation stability of the weir foundation.

4.9.3 Make preliminary engineering geological classification of surrounding rock of the diversion tunnel. Evaluate the slope stability of open diversion channel.

4.10 Engineering Geology along Water Conveyance Routes

4.10.1 Brief engineering geology along water conveyance route.

4.10.2 Evaluate the major engineering geological problems of alternative schemes. Provide opinions on schemes selection from geological point of view.

4.10.3 Basically determine main physical and mechanical parameters for rock mass and soil. Evaluate the major engineering geological problems of the recommended route for the main structures.

4.10.4 Carry out engineering geological sectioning for channels of the recommended route. Evaluate major geological problems for each section of the channel.

4.10.5 For tunnels of the recommended route, evaluate the major geological problems of surrounding rock and the slopes of inlet and outlet. Make preliminary engineering geological classification of the surrounding rock.

4.11 Engineering Geology of Levee and River Training Works

4.11.1 Brief engineering geology of levee and river training works.

4.11.2 Evaluate the major engineering geological problems of the schemes of levee alignment. Provide engineering geological opinions on schemes selection. Basically determinate the main physical and mechanical parameters for rock mass and soil.

4.11.3 Evaluate the major engineering geological problems of levee foundation in sections. Preliminarily evaluate the levee embankment quality and problems/defects.

4.11.4 Evaluate the major engineering geological problems of appurtenant structures such as culvert and sluice.

4.11.5 Evaluate stability of river revetment in sections and propose soil excavation classification and excavation slope ratio.

4.12 Hydrogeology of Irrigation Area

4.12.1 Brief regional geology and hydrogeology, including topography and geomorphy, stratigraphy and lithology, tectonics, main aquifers and their recharge/discharge relations.

4.12.2 Brief hydrogeology of ground water resource area for irrigation. Determine hydrogeological parameters. Calculate and evaluate the allowable amount of groundwater extraction and conduct reliability analysis.

4.12.3 Brief hydrogeology of the irrigation area. Predict possible changes of ground water table and quality, and subsequent impacts on the land use—the probability of salinization and bogginess due to the project operation. Propose remedial measures. For the existing irrigation (discharge) area, evaluate the property, distribution and situation of the salinized soil.

4.13 Natural Construction Materials

4.13.1 Indicate requirements of types, quantity and quality of natural construction materials required for the project.

4.13.2 Brief location, topography and geomorphology, geology, exploitation and transportation conditions of each borrow area/quarry.

4.13.3 Evaluate the reserve and quality of materials in each selected site.

4.14 Figures and Tables

4.14.1 Following figures may be attached to this chapter:

1 Regional geological map (with stratigraphic column) or regional tectonic sketch map and seismic epicenter distribution map.

2 General geological map of the reservoir area.

3 Engineering geological map of schemes.

4 Engineering geological map and transversal and longitudinal profiles of main structures.

5 Hydrogeological map of irrigation area.

6 Distribution of borrow area/quarry of natural construction materials.

4.14.2 Following tables may be attached to this chapter:

1 Summary of completed site investigation work.

2 Summary of test results.

5 Purposes and Scale of Project

5.1 Necessities and Purposes of Project Development

5.1.1 Brief the situation in connection with project location, socioeconomic status, and conditions of nature, geography and resources, status quo of water and hydropower projects, and other important projects in the region where the project is located.

5.1.2 Brief development plan of the river where the project is located and main conclusions of relevant review, as well as the scheme, results and approval comments in the project proposal stage.

5.1.3 Elaborate the requirements of national socioeconomic development for water and hydropower project in the region where the project is located, the implementation schedule for the proposed project involved in short and long term development plan of water and hydropower, and the roles and functions of the proposed project to the local area in national socioeconomic development, as well as in the overall arrangement of river training and development.

5.1.4 Assess the necessity of the project from following aspects:

1 For flood (tide, ice run) control projects, elaborate the historical major disasters happened in the project area such as flood, storm surge and ice run as well as their damages and impacts on local economy and society, the current status of flood (tide, ice run) control facilities and problems existed, and the requirements of socioeconomic development on improving the capability of flood (tide, ice run) management. Analyze the function and benefits of the project in preventing disasters and mitigating risks and its role in safeguarding local socioeconomic development.

2 For waterlogging control projects, elaborate the historical major waterlogging disasters, their damages and impacts on local economy and society, the current status of stagnant water drainage facilities and problems existed, and the requirements of socioeconomic development on improving the capability of stagnant water drainage. Analyze the benefits of the project in waterlogging control and risk mitigation and its role in local socioeconomic development.

3 For river channel and estuary training works, elaborate the situation of local river channel and estuary evolution, change in aggradation and degradation, the impacts of socioeconomic development, human being activities, variation of water and sediment characteristics on river channel and estuary. Elaborate the existing status and problems of river channel and estuary training works, the characteristics of water and sediment, the trend of river channel and estuary evolution, and the requirements of local socioeconomic development on river channel and estuary training and project implementation. Analyze the function and benefits of the project in local flood (tide, ice run) management, water logging control, agriculture and land utilization, aquatic cultivation, navigable channel training, water resources utilization and ecological protection. If there are reclamation requirements in river channel and estuary areas, elaborate the status of reclamation areas, the development and requirements of reclama-

tion, shoreline and estuary development plan, and analyze the requirements of local socioeconomic development, shoreline utilization and reclamation on water resources and hydraulic works.

4 For irrigation projects, elaborate the status of agriculture and livestock farming in the irrigation area, the situation and characteristics of main droughts and salinization and their impacts on agriculture and livestock farming, and elaborate status and problems existed in irrigation facilities, drinking water of human and livestock and water resources in the irrigation areas. Analyze the requirements of local agriculture and livestock farming and grain production on irrigation and drainage, and the function and benefits of the project development for local socioeconomic development, multipurpose development of agriculture and livestock farming and grain production.

5 For water supply projects, elaborate the status of water resources and water quality in water receiving area, the situation of water consumption, water supply and water saving for urban living, industry, countryside and environment, problems existed in water resources development and utilization, elaborate the influences and restrictions of water shortage on local socioeconomic development and people's livelihood, and the requirements of socioeconomic development on optimization of water configuration, development and utilization of water resources, construction of water supply projects. Analyze the function and benefits of the project in socioeconomic development and water supply safety of the water receiving areas. For ecological and environmental protection water supply, elaborate local ecological and environmental situation and tendency, analyze the impacts of the project on ecology and environment of local and surrounding areas, and describe the importance and necessity of protecting, restoring and improving the ecology and environment.

6 For hydropower projects, elaborate the situation of local energy sources, the status and problems of power industry, power system development plan and power demand. Analyze the demands of local socioeconomic development on hydropower resources exploitation. Brief the role and function of the project in power system and river cascade development. Analyze its role in accelerating local economic development and power system construction.

5.1.5 Elaborate the function and benefits of the project in safeguarding social harmony and stability, improving ecological environment and navigation, preventing soil erosion, and promoting tourism development and aquaculture.

5.1.6 Analyze the requirements of local socioeconomic development and various sectors on multipurpose utilization of the project. Determine purposes of the project and their priorities in combination with the project conditions and coordinating the requirements of various sectors.

5.2 Flood Control

5.2.1 Brief the local socioeconomic conditions and development plan, the characteristics of flood occurrence in the river basin, the observed and historical floods, flood disaster, and status and requirements of flood control.

5.2.2 Analyze the requirements of protection objects, and determine the protection scope, objects and standard of flood control.

5.2.3 Describe the formation of regional flood, flood control engineering system, functions of main flood control projects and the principle of flood regulation.

5.2.4 Determine the general layout of flood control projects, contents of construction and engineering measures.

5.2.5 Justify the scale of levee projects from following aspects:

1 Describe the protection scope and objectives of the levee project.

2 Brief the historical evolution of the levee, form of cross-section, dangerous structures and sections, number and types of structures crossing over/through the levee and damages of the structures, and the relationship between river crossing structures and the levee.

3 Determine the scope, principle and design standards of levee reinforcement.

4 Determine the safety discharge capacity of various reaches and design water levels at various control sections, review the flood discharge capacity of river with the existing levee.

5 Select the levee alignment and spacing of levees, basically select the cross-section type of flood passage, and calculate the water surface profile. For the newly constructed river channel for flood diversion and the newly constructed levees, select the alignment and spacing according to the river channel geomorphology, flood discharge requirements, land occupation on river banks and environmental impacts after comprehensive technical and economic comparison.

6 Basically select the size of levee, and the location and size of river (levee) crossing structures.

7 Analyze the flood discharge way and size of main channel and floodplain area, basically determine the protection measures for the floodplain area.

8 Basically determine the area, objects and measures of river channel clearing.

9 For the tide affected river reach, analyze the impact of tidal level on flood discharge.

10 For river channel where ice run control is needed, analyze the characteristics of ice run, basically determine the discharge and water level of ice run control, comprehensively consider the requirements on flood and ice run control, basically determine the size of ice run control project, and propose the operation mode of ice run control.

5.2.6 Justify the scale of flood storage and detention project from following aspects:

1 Describe the purposes, area, criteria for putting into use, and operation conditions of the flood storage and detention project.

2 Determine the area, design water level, storage capacity, flood diversion level and discharge of the flood storage and detention project.

3 Determine the general layout of water retaining structures, water intaking structures, water evacuating structures and connection works in the flood storage and detention project.

4 Determine the size and main parameters of backbone projects.

5 Develop regulation principles and modes of the flood storage and detention project, and provide water levels at different controlled sections based on the flood regulation and backwater calculations.

6 Preliminarily determine the development plan of safety and management of the flood storage and detention project.

5.2.7 Justify the scale of reservoir project from following aspects:

1 Describe requirements of downstream area on flood control, determine the discharge capacity or flood control water level at flood control cross-sections.

2 Basically determine the maximum allowable discharge and the required storage for flood control.

3 Basically determine the flood regulation principles of the reservoir. For cascade reservoirs, basically determine integrated regulation principles for flood control of cascade reservoirs.

4 Describe the basic data and calculation methods of flood routing.

5 Describe the principles and range for selecting the limited water level in flood season; analyze the inundation area of reservoir, important inundated objects, major constraints, difficulty in resettlement, construction conditions and geological conditions. Determine the limited water level in flood season through comprehensive technical and economic comparison. Basically determine the maximum water level for flood routing, size of water releasing structures and other characteristic water levels.

6 If necessary, analyze the phased flood control levels of reservoir, and basically select the phased limited water level for flood control in flood season.

7 For the reservoir with requirements of ice run control, analyze the ice run characteristics of the reservoir area and river channel and the requirements of ice run control, and basically select the ice run control storage, and propose regulation mode of ice run control.

5.2.8 Analyze the effectiveness and influence of reservoir flood control.

5.3 Waterlogging Control

5.3.1 Brief the situation of physical geography, socioeconomic development, local development plan, characteristics of river system and geomorphology in the waterlogging area. Analyze the rainfall, characteristics of waterlogging, scope of and causes of waterlogging disasters, and situation and existing problems of waterlogging control project.

5.3.2 Describe the local socioeconomic development plan, the water resources development plan, and the requirements of waterlogging control. Determine the scope of waterlogging control area, and basically determine the zoning of waterlogging control area.

5.3.3 Analyze the water drainage requirements of protective objects, benefit and cost of the waterlogging control project. Determine standards and principles of waterlogging control. Harmonize the waterlogging control standards of overall and partial, main stream and tributaries.

5.3.4 Analyze the cases of local flood and waterlogging concurrence, determine the regulation principles of flood and waterlogging control, as well as the drainage river channel, the drainage receiving area, the waterlogging detention area, and the way of dewatering, and select the general layout and main works of waterlogging control project.

5.3.5 Analyze the waterlogging drainage modulus. Determine the waterlogging drainage discharge, and basically determine the design water level and discharge of main drainage channels (ditches), culverts, sluices and levees, the water level and storage of waterlogging detention area, the water level of drainage receiving area, and the design water level, discharge, installed capacity and main parameters of waterlogging drainage pump stations.

5.3.6 Basically determine the regulation mode of waterlogging control project.

5.3.7 For waterlogging area with large range and many subzones, a representative and typical zone may be selected to conduct the design of waterlogging control project.

5.4 River Channel and Estuary Training

5.4.1 Brief the situation of local physical geography, socioeconomic development of the area, and the status and training plan of the river channel and estuary. For a navigable river channel, clarify the status of navigation and relevant plans, and describe the characteristics of local storm, flood, typhoon, tide, sediment and the topographic and geological conditions.

5.4.2 Determine the training scope of river channel and estuary.

5.4.3 Based on the importance of the river reach to be trained and the training situation upstream and downstream of the reach, determine the training criteria through technical and economic comparison. Calculate and determine the river bed-forming discharge, and select the design water (tide) level, discharge and width of the reach.

5.4.4 Describe the hydrological and sediment characteristics. Analyze the evolution law of the river channel and floodplain and the estuary, the evolution trends of river regime; and select the training alignment.

5.4.5 Comprehensively analyze the requirements for riverside utilization, land development, flood control and risk mitigation, utilization of water resources, etc. Select general layout of the river channel and estuary training, location of controlled nodes and water (tide) retaining levee line of important reaches, main works and engineering measures through technical and economic comparison.

Selection of water (tide) retaining levee alignment shall meet the overall training requirements. It shall conduct specific study and physical model test on river channel if the flow and tide pattern is complex or the training works have significant impact on river regime.

5.4.6 Determine the sizes and main parameters of water control and diversion works, river channel dredging, floodplain and shoal training, levee (coastal levee) construction and sluice of flood (tide) retaining at estuary. Analyze the impacts of river course dredging and levee construction on river (levee) crossing structures, and basically select the reconstruction scheme of structures.

5.4.7 For flood (tide) retaining sluice at estuary, analyze river runoff, flow (tide) downstream of sluice, river regime, sediment scouring and sedimentation characteristics at estuary before and after sluice construction, analyze measures of deposition prevention and operation mode, and propose water volume for sediment flushing.

5.4.8 For land reclamation of river channel and estuary, describe the requirements of land utilization, industrial and agricultural production, aquiculture, etc. Determine scope and area of reclamation area, design standards of flood control, tide retaining, water diversion and water drainage; determine the general layout and main construction works for protection of reclamation area and water supply and drainage project. Select flood (tide) retaining levee line, design flood level, flood (tide) retaining level, sizes and main design parameters of main flood (tide) retaining levee, embankment, culvert and sluice.

5.5 Irrigation

5.5.1 Brief the situation of local geography, socioeconomic development in the irrigation area and area where the irrigation project is located, the status and development plan of land use, irrigation facilities, drinking water for human being and livestock, agriculture and livestock farming in the irrigation area.

5.5.2 Determine the scope of irrigation area, and basically select the irrigation method.

5.5.3 Investigate the status of land use in the irrigation area. Analyze the requirements of local socioeconomic development, status of water and soil resources, and the requirements of multi-purpose development. Determine the classification of irrigating land and land use plan, analyze soil types and conditions of development and utilization, and select the irrigation area through comprehensive technical and economic comparison.

5.5.4 Determine design reference year, design dependability of irrigation and drinking water for human being and livestock in the irrigation area.

5.5.5 Basically determine the production structure of agriculture, forestry and livestock as well as crop types. Analyze the irrigation modes, water consumption, water demand of various crops, and water demand and quota of drinking water for human being and livestock in the irrigation area. Basically determine the utilization coefficient of irrigation water. Propose total water demand and annual distribution in the irrigation area.

5.5.6 Analyze available water for different water sources and different years, balance water supply/demand in the irrigation area, and propose the amount of water demand/supply and water supply hydrograph.

5.5.7 Select the general layout of irrigation projects, main works and staged implementation plan.

5.5.8 Justify the scale of water source works from following aspects:

1 Describe the hydrological runoff characteristics of the river, and water use plan upstream and downstream of the project.

2 For reservoir projects, basically select the active storage that meets the water demand of irrigation. Analyze the probable range of normal pool level. Select the normal pool level through comprehensive technical and economic comparison. Analyze the land distribution in the irrigation area, conditions of gravity flow for irrigation, and sedimentation of reservoir. Basically select the dead water level and other characteristic water levels.

3 For water conveyance project and pump station taking water directly from river channel, statistically analyze the assurance level of runoff in the river, and select the design diversion discharge and design water level.

4 For irrigation area with groundwater source, analyze the recharge relation between regional surface water and groundwater, the exploitable yield of ground water. Basically determine the groundwater supply volume.

5 Present the basic data and describe the method of runoff regulating simulation. Analyze the simulation results of long runoff series. Provide the indices such as design diversion discharge, average annual water diversion, probability of irrigation, etc.

5.5.9 Justify the scale of irrigation canals from following aspects:

1 Basically determine the routes of main canals and distributaries, and positions of structures of canal system.

2 Select the design discharge of main canals of irrigation, and basically determine the design water levels at main nodes.

3 Basically select the size and main parameters of canal –

crossing structures, pump stations, storage reservoir in irrigation area, sedimentation treatment and water releasing structures.

5.5.10 Analyze possible causes of salinization in the irrigation area; propose the zoning for soil improvement considering the topography, soil, hydrogeology, and technical and economic conditions of the irrigation area. Basically select the layout, scale and drainage pattern of water drainage projects.

5.5.11 Basically determine the irrigation design and scale of irrigation and drainage canal systems in a typical farmland.

5.5.12 Analyze impacts of water abstraction of water source works for the irrigation area on other water users, and propose remedial measures if necessary.

5.6 Water Supply

5.6.1 Brief the total volume and distribution of water resources in water receiving area, present the status and existing problems of water resource exploitation, and water supply conditions of urban and rural water supply and existing water source projects.

5.6.2 Describe the socioeconomic development plan and water demand in water receiving area. Determine the scope of water supply, main water supply objectives, water supply objects, design reference years, and design dependability of water supply for different water users.

5.6.3 Describe the socioeconomic development plan and industrial structure in water receiving area. Determine water saving target and propose measures. Analyze water use quota of industrial sectors in different design reference years, and quota of municipal and domestic water use.

5.6.4 Describe the requirements of various water users on water volume, flow and water quality. Predict the water demand of various sectors, total water demand and hydrography in different design reference years.

5.6.5 Analyze the planning water sources and available water supply in different design reference years in water receiving area. Estimate the utilization volume of recycled water based on water saving and pollution control plan. Describe the allocation and utilization principles of local water resources in the water receiving area. Analyze the balance of water demand and supply in the water receiving area, and propose the water shortage volume, amount of water demand/supply and water supply hydrograph.

5.6.6 Describe the allocation principles and schemes of regional water resources, and assess the available water volume and reliability of water supply for the proposed water source projects in different design reference years. For inter − basin or inter − region water transfer projects, describe the status of water resources development and utilization, socioeconomic development plan in the water supply area, analyze water use of various sectors and water demand of ecological environmental flow in the water supply area, and determine the available water volume to be taken from the water supply area.

5.6.7 Select the general layout and main works of water supply projects through comprehensive technical and economic comparison.

5.6.8 Justify the scale of water source and storage and regulation works from following aspects:

 1 Describe hydrological runoff characteristics of the river, and planned water use both upstream and downstream of the project.

 2 For reservoir project, analyze the active storage that meets the water supply demand, analyze the probable range of normal pool level, and select the normal pool level through comprehensive technical and economic comparison. Describe sedimentation situation and the requirements of water conveyance project layout on the elevation of water intake, and basically select the dead water level and other characteristic water levels.

3 In case of reservoir project with large dead storage, important water users, and high water supply probability, analyze the measures of utilizing the dead storage for emergency water supply during extraordinary dry year, water supply volume and corresponding minimum water supply level if necessary. Propose the requirements on layout and elevation of intake.

4 Based on the status of water use both upstream and downstream of reservoir, and predicted water demand in the design reference year, analyze the reliability of water supply from reservoir.

5 For water conveyance project or pump station taking water directly from river channel, analyze the assurance level of runoff in the river, and select the design diversion discharge and design water level.

6 For water conveyance project taking water from tide affected river reach, analyze the impact of tide (salt water) on the water intake, and determine the design water intake period and discharge.

7 For area with groundwater sources, analyze the recharge relation between regional surface water and groundwater, the exploitable yield of groundwater; basically determine the groundwater supply volume.

5.6.9 Justify the scale of water conveyance project from following aspects:

1 Select the design discharge of main water conveyance canals (pipelines, box culverts, and tunnels).

2 Basically select the design water levels at main nodes.

3 Basically select the sizes and parameters of water intake gate, crossing structures, pump stations and so on.

5.6.10 Comprehensively analyze the technical and economic conditions of water receiving area, water providing area and water supply projects. Carry out the regulating simulation of long series, determine the design annual water supply volume and average annual water supply volume in water receiving area. Describe the water loss in water conveyance route, and propose the annual water diversion of the water supply project.

5.6.11 Analyze the sediment characteristics if water is taken from sediment – laden river, and basically determine the layout and scale of sediment treatment works.

5.6.12 Analyze the change of discharge in the river before and after water diversion, and impacts of water diversion on water use for production activities, livelihood and ecological environment inside and outside of the river in water providing area. Propose treatment measures if necessary.

5.6.13 Describe the status of water quality and quantity of the reservoir, and propose the requirements and measures for water source protection.

5.6.14 Basically determine the water allocation principle, and propose integrated regulation scheme for the transferred water and local water resources.

5.7 Hydropower

5.7.1 Describe the situation of regional power resources, including water power, coal, petroleum, natural gas and so on. Describe the socioeconomic characteristics, and the power demand of regional economic development. Analyze the potential power supply scope of the hydropower project, and its purpose and role in power system.

5.7.2 Describe the status of power supply and demand, and power development plan in power supply area. Analyze the trend of regional electrical load development and the characteristics of power system operation. Determine the design reference year, forecast the load level, analyze the load behavior, and basically determine the design dependability of power generation.

5.7.3 Present the basic data and describe the method of energy estimation. Propose the operation mode of reservoir and hydropower project.

5.7.4 Justify the characteristic water levels of hydropower project from following aspects:

1 Describe the selected characteristic water levels upstream and downstream of the hydropower project at project proposal stage. Analyze the cascade linking water level, inundation area of reservoir and constraints, engineering geology and conditions for project layout. Select the normal pool level of the reservoir through comprehensive technical and economic comparison.

2 Analyze the impact of sedimentation elevation and turbine-generator unit operation characteristics on drawdown and minimum operation water level of the reservoir, propose the range of dead water level, and basically select the dead water level and other characteristic water levels of the reservoir.

5.7.5 Determine the design diversion discharge of diversion-type hydropower project and design water level of power intake and forebay.

5.7.6 Select the installed capacity of hydropower project from following aspects:

1 Brief the development situations, load prediction, load characteristics, composition of power sources, development plan of power system, and the characteristics of other existing, under-construction and to-be-constructed hydropower projects.

2 Describe regulation characteristics of the reservoir, energy economic indices of hydropower station, and added benefit to downstream cascade hydropower project by this project. Analyze the alternative schemes of installed capacity and conditions of project layout. Select the installed capacity through comprehensive technical and economic comparison.

3 Propose the procedure of putting units into operation and opinions for reserved places for units in future for the hydropower projects with long time of initial operation stage or with a big change in short or long term energy index.

4 Brief characteristics of the runoff and reservoir operation, analyze characteristic water heads and technical and economic indices of hydropower project, and basically select the rated head.

5.7.7 Calculate indices such as firm output and average annual output of hydropower project, and analyze the results of runoff regulating simulation.

5.7.8 Analyze impacts of the water intaking and the discharge for peak load on water use within and outside of the downstream river and ecological environment, and propose mitigation measures if necessary.

5.8 Multi-Purpose Project

5.8.1 Brief the socioeconomic status, nature, resources and environment of the area where the project is located.

5.8.2 Describe the requirements of multi-purpose sectors on the reservoir project, such as flood control, irrigation, water supply, power generation, aquatic environment protection, navigation, tourism, and prioritize the purposes of reservoir by considering the project conditions.

5.8.3 For multi-purpose project, coordinate the relationship of all purposes. Justify the main scale indices including the reservoir storage capacity and characteristic water levels to meet the requirements of each main purpose and taking other purposes into account. Select the normal pool level and limited water level in flood season of the reservoir. Basically determine the active storage and flood control storage, the dead water level and other characteristic water levels and installed capacity. Prepare special study report on normal pool level comparison if necessary.

5.8.4 Analyze active storage and flood control storage of the reservoir from following aspects:

   1 For the active storage, describe the water volume, flow rate and water level of the reservoir required for water supply, irrigation, power generation and navigation. Analyze the water use characteristics of various water users and the ecological flow in the river. Coordinate water demand and relationship of various water users. Propose the active storage of the multi-purpose reservoir.

   2 For the flood control storage, describe downstream flood control requirements, operation mode and storage required for flood control. Propose phased flood control storage if necessary.

5.8.5 Select normal pool level of the reservoir from following aspects:

   1 Describe the cascade linking water level proposed in the project planning stage and the normal pool level scheme selected in the project proposal stage.

2 Describe the basic data, hydrological runoff results, runoff regulating simulation period, upstream and downstream water using process, and principles and methods for regulating simulation.

3 Propose the selection principles, range and schemes of normal pool level. Analyze the inundation area of reservoir and important inundated objects and major constrains, construction conditions and geological conditions. Select normal pool level of the reservoir through comprehensive technical and economic comparison.

4 For a reservoir project of staged development, justify normal pool level and other characteristic water levels for the initial stage and the final stage separately.

5.8.6 Select dead water level of reservoir from following aspects:

1 Describe the dead water level scheme selected at the project proposal stage.

2 Comprehensively analyze the situation of reservoir sedimentation, and requirements of water supply, irrigation, power generation, fisheries, navigation and reservoir sediment scouring for the minimum operating level of the reservoir. Propose the range of dead water level and select the dead water level of reservoir in consideration of the technical and economic factors.

3 Justify the adoption of different dead water levels for short term and long term if the change in sedimentation and water use is great.

4 Preliminarily propose emergency measures of water supply, and select the minimum operating level if necessary.

5.8.7 Conduct flood routing and select flood control characteristic water levels from following aspects:

1 Describe the limited water level scheme in flood season selected at the project proposal stage.

2 Describe the basic data for flood routing, flood in various return periods, typical flood hydrograph and phased floods.

3 Propose the selection principles of limited water level in flood season. Coordinate the relationship among flood control, water supply and power generation. Analyze the factors such as inundation area of reservoir, expected objectives and benefits, construction conditions, geological conditions, etc. Select the limited water level in flood season, flood control storage, the maximum water level for flood routing, design flood level, check flood level and scale size of water releasing structures through comprehensive technical and economic comparison.

4 Study the limited water level in flood season for phased flood control, and analyze the benefits of reservoir water supply and the phased flood routing results if necessary.

5 Analyze the impact on flood control of the project due to upstream dam failure.

5.8.8 Select installed capacity of the hydropower project from following aspects:

1 Describe the basic data, basic method and conditions of energy estimation, and analyze the relationship among water use for power generation, water supply, irrigation, and river ecological environment.

2 Justify the scale of hydropower project and select the installed capacity according to the multi-purpose requirements of water supply and relevant contents and requirements in Section 5.7.

5.8.9 Analyze reservoir sediment scouring and deposition from following aspects:

1 Brief the sediment characteristics, describe the characteristic value of design sediment series and distribution within a year, and the principle, method and main parameters of sediment calculation.

2 Analyze objectives, reservoir regulation performance, reservoir morphology, and water and sediment characteristics of the multi-purpose project. Propose the sediment scouring and deposition calculation results of the reservoir in various design reference years.

3 Provide the measures on maintaining long-term active storage capacity of reservoir, reducing the sediment deposition and inundation at the end of reservoir, and operation schemes of water and sediment regulation.

4 Basically determine the operation modes and provide measures on sediment controlling and desilting at the intake of water diversion structures.

5 Analyze the impact of reservoir sedimentation on the tail water of the upstream cascade hydropower project and the flood control of both banks of the river.

6 Analyze the impact of the downstream reservoir, river channel on operation modes of flood releasing and sediment scouring of the proposed reservoir.

5.8.10 Carry out the calculation and analysis of reservoir backwater from following aspects:

1 Describe the basic data available, calculating conditions and methods, and corresponding flood standards for different inundated objects in the reservoir area.

2 Calculate the reservoir backwater surface profile, and determine the pinchout point of backwater comparing with the natural water surface profile of the same return period.

3 For the reservoir subject to obvious sedimentation impact, propose the sedimentation distribution with different deposition years in the reservoir area, and analyze the impact on backwater.

4 For the areas with heavy ice run, analyze the ice run characteristics and conditions of ice dam formation in river channel and at the end of reservoir, and the water level rising due to ice barrage and the scope of backwater, and propose the treatment measures.

5.8.11 Determine reservoir operation mode and annual operation characteristics with following aspects:

1 Analyze the present water use of upstream area of the reservoir and the predicted water consumption in design reference year, and analyze the water supply dependability of reservoir water source.

2 Describe the multi-purpose requirements of reservoir, reservoir regulation principles and scheme, and integrated regulation mode with other projects. Preliminarily plot reservoir operation diagrams for important projects.

3 Propose indices of the average annual water supply, water supply in different representative years, water supply to different objects, probability of water supply, firm output and average annual energy, etc.

4 Analyze the extent of water use, the extent in fulfilling the requirements of various sectors and benefit indices. For storage capacity served both as active storage and flood control storage, analyze the reservoir impounding situation after flood season. For main regulation reservoirs, propose the results of compensation regulation of cascades or inter-basins if necessary.

5 Describe the variation of downstream river discharge and water level before and after reservoir construction. Analyze the impact of reservoir operation on downstream water demand and river regime, river channel erosion and works along the river banks. Propose the prevention and treatment measures if necessary.

6 For the reservoir with large storage and great regulating capacity, preliminarily propose plans of initial impoundment, water supply, and power generation, analyze potential impact of initial impoundment on upstream and downstream of the reservoir.

5.8.12 For other projects with multi-purpose utilization and comprehensive regulation purposes, basically determine the operation scheme, select the size of main structures, and basically determine the size of other structures.

5.8.13 For a multi-purpose project with navigation requirements, determine the navigation criteria, predict the passenger and freight volume passing the dam (sluice) in design reference year. Basically determine upstream and downstream navigable water levels and discharges in the project area and determine the size of navigation structures.

5.9 Reinforcement, Renovation and Extension Projects

5.9.1 For reinforcement projects, describe original design scale of the project, existing problems in the operation and recent safety inspection and assessment conclusion. According to the current operating condition, recheck whether the project fulfills the original functions and assess the project safety, justify necessity of the reinforcement project, and basically determine the scope of works.

5.9.2 For renovation and extension projects, describe original design scale of the project and existing problems in the operation, justify the necessity and benefits of the renovation and extension projects, and the impact on the environment and other water users. Basically determine the scope and scale of the works.

5.10　Figures, Tables and Appendices

5.10.1　The following figures may be attached to this chapter:

　1　Schematic diagram of river basin (reach) and regional multi-purpose plan.

　2　Schematic diagram of general plan.

　3　Risk chart of flood protection area.

　4　General layout of the project.

　5　Diagram of balance of electric output and energy.

　6　Chart of reservoir stage - volume - area curve (original and after sediment deposition).

　7　Schematic diagram of reservoir flood regulation process.

　8　Longitudinal profile of the reservoir backwater (original and after sediment deposition).

　9　Plan of sections for backwater calculation.

　10　Other figures.

5.10.2　The following tables may be attached to this chapter:

　1　Analysis results of water resources supply and demand balance.

　2　Results of design water surface profile.

　3　Scheme comparison of reservoir normal pool level.

　4　Schemes comparison of limited water level in flood season.

5 Results of runoff regulating simulation.

6 Schemes comparison of installed capacity of hydropower project.

7 Calculation results of reservoir (sluice) backwater.

8 Others.

5.10.3 The following reports may be attached to this chapter:

1 Project scale justification.

2 Irrigation area plan.

3 Analysis on supply and demand of water resources.

4 Analysis on sediment scouring and deposition.

5 Others.

6 Project Layout and Structures

6.1 Design Basis

6.1.1 Brief the main review and approval comments on and conclusions of the project proposal report.

6.1.2 Describe the basic data and information in related disciplines needed for selection of project site and general layout, and design of main structures.

6.1.3 Describe the main technical standards for design.

6.2 Project Grade and Standard

6.2.1 Describe the project scale, scope of works and main structures. Determine the project grade, structure class and corresponding flood design standard based on analysis and justification.

6.2.2 Determine the design values of seismic parameters and corresponding design seismic intensity.

6.2.3 Describe the main allowable design values specified in prevailing national technical standards/codes.

6.3 Selection of Project Site and Route

6.3.1 Describe the principles of comparison and selection. Propose the potential project sites or routes as candidates after analysis based on the conclusion of the project proposal.

6.3.2 Conduct comprehensive study on and comparison of the alternatives of the sites of dam, sluice, pump station, powerhouse, etc. and select the project site.

6.3.3 Select the levee alignment based on the comprehensive justification and comparison of the alternatives.

6.3.4 Select the water conveyance route based on the comprehensive justification and comparison of the alternatives of irrigation, drainage and water supply pipeline.

6.3.5 For main structures with very complicated geological conditions and particularly difficult construction conditions, conduct special study on the site and route selection.

6.4 Selection of Structural Types

6.4.1 Based on comprehensive justification and comparison, select basic dam type, and basically select the representative dam type, and structural types of water releasing structures, diversion structures, dam – passing – over structures, etc.

6.4.2 Based on comprehensive justification and comparison, basically select the main structure type of the sluice, pump station, hydropower station, etc.

6.4.3 Based on comprehensive justification and comparison, basically select types of the levee and main levee crossing structures.

6.4.4 Based on comprehensive justification and comparison, basically select water conveyance manners of the irrigation, drainage and water supply projects and types of main structures.

6.4.5 Based on comprehensive justification and comparison, basically select the types of river training works or man – made river channel structures.

6.4.6 Based on requirements of the layout of main structures, preliminarily select the type of secondary structures.

6.4.7 Conduct special study when structures are constructed with new type of materials or in new structural types, or in case of insufficient experience.

6.5 General Project Layout

6.5.1 According to the selected project site and type of main structures, basically select the general layout of reservoir dam, sluice, pump station, hydropower station, etc., subject to comprehensive justification and comparison.

6.5.2 Based on the existing status of the levee and river training works and the requirements on flood control, waterlogging control, navigation and beach land utilization, basically select the general layout of levee and river training works.

6.5.3 Based on comprehensive justification and comparison, basically select the general layout of irrigation, drainage and water supply projects, and basically select the structural type and layout of canal network, crossing structures and appurtenant structures.

6.5.4 Preliminarily select the design scheme of architectural style and architectural esthetics of the project area and main structures.

6.5.5 Summarize the number and main characteristic indices of various structures.

6.6 Water Retaining Structures

6.6.1 Basically select the layout of water retaining structures, including main control elevations, structural type, dimensions of main structures, and means of connection with abutments or other structures. Preliminarily carry out the structural design.

6.6.2 Propose the requirements for concrete bulk density, concrete zoning of dam body, concrete strength, impermeability and freezing resistance grade, and raw materials. Propose design requirements and compaction standard of materials in different zones of earth and rock fill dam. Propose requirements for masonry, geo-synthetic materials and other materials.

6.6.3 Preliminarily determine the requirements for foundation excavation, seepage control, reinforcement measures including the treatment for foundation with poor geological conditions.

6.6.4 Describe the calculation condition, physical and mechanical characteristic parameters, loads and load combinations, calculation method and preliminary results for the stability, stress, deformation, leakage, seepage of structures and their foundation. Propose the criteria of temperature control for concrete dam. For important projects or structures with complicated shapes, special study on temperature control should be carried out.

6.7 Water Releasing Structures

6.7.1 According to preliminarily selected type and general layout of water releasing structures, basically select the configuration, structural type, main control elevations, structural dimensions and energy dissipation type. For the water releasing structures with high velocity flow and with the functions of sediment scouring, trash skimming, ice and floating debris releasing, propose corresponding structure arrangement and measures against cavitation, abrasion and scouring.

6.7.2 Propose the requirements for concrete bulk density, concrete zoning, concrete strength, impermeability and freezing resistance grade, and concrete materials.

6.7.3 Preliminarily determine the requirements for foundation excavation and measures for seepage control and foundation reinforcement.

6.7.4 Preliminarily select the rock support, type of lining and surrounding rock reinforcement measures of water releasing tunnels. Preliminarily select the scope and measures of grouting, seepage control and drainage.

6.7.5 Describe the hydraulic conditions of water releasing structures, such as operation mode, discharge capacity, flow pattern, hydraulic gradient line, energy dissipation, hydraulic connection in upstream and downstream areas, sediment scouring, aggradation and degradation of downstream river and its impacts. Describe the conditions and approaches of hydraulic calculation, and propose preliminary calculation results. For water releasing structures of important project, essential hydraulic model tests should be conducted. For water releasing structures of important projects on heavy sediment–laden rivers, essential sediment tests or mathematical model analysis should be conducted.

6.7.6 Describe the preliminary calculation results of stability, stresses and deformations of main structures. Describe the calculation condition, loads and load combination, approach and preliminary calculation results for tunnel lining. For important projects or structures with complicated shape, special study should be carried out.

6.8 Diversion Structures for Power Generation

6.8.1 Basically select the layout, structural type, main control elevations and structural dimensions according to the selected type of diversion structures.

6.8.2 Describe conditions and methods of hydraulic computations of diversion structures, and propose the preliminary calculation results.

6.8.3 Justify the necessity of surge chambers/shaft, and basically select the layout scheme, structural type, control elevations and main dimensions of the surge chambers/shaft.

6.8.4 Describe conditions and methods for stability and structural calculation of the structures, and propose preliminary calculation results.

6.8.5 Preliminarily select the foundation treatment measures for diversion structure.

6.9 Powerhouse (Pump House) and Switchyard (Substation)

6.9.1 Basically select the layout, structural type, area, elevations and main dimensions of main powerhouse and auxiliary powerhouse. Describe conditions, methods and preliminary results of the stability and structural calculations. Basically determine the layout, type, control elevation and main dimensions of tail water (effluent) structures.

6.9.2 Provide the results of stability analysis on surrounding rocks of caverns of underground powerhouse, and basically select the supporting and lining types.

6.9.3 Basically select the foundation treatment measures for powerhouse/pump house.

6.9.4 Basically select the layout, structural type, area, elevation and main dimensions of the switchyard (substation), as well as the foundation treatment measures for the switchyard (substation).

6.10 Slope Works

6.10.1 Preliminarily determine the position and scope, the class and design criteria for slope works.

6.10.2 Propose the calculation methods and results of slope stability analysis, and evaluate the stability of the slope. Preliminarily select the gradient of slope excavation and step height of single slope, berms, position and dimensions of platforms. For slopes to be treated, basically determine the design principles of slope treatment, and preliminarily select the treatment measures.

6. 10. 3 For slopes whose deformation needs to be controlled, preliminarily select technical measures for controlling the deformation.

6. 10. 4 For landslide, imminent falling and collapsing rock mass, preliminarily select treatment measures based on the type of landslide, inducing factors and characteristics of sliding mass.

6. 11　Irrigation and Drainage Structures

6. 11. 1 Based on comprehensive comparison and selection, preliminarily select the layout, structural type, control elevations and main dimensions of such structures as head works, main canal system, and water storage and regulation works. Summarize the number and main characteristic indices of the structures.

6. 11. 2 Describe computation conditions and methods of flow rate of the irrigation and drainage works, describe the calculation conditions and methods of water level, hydraulic gradient line of the headrace, irrigation and drainage works, and propose primary hydraulic calculation results. Describe the calculation conditions and methods of the discharge capacity, hydraulic connection, energy dissipation and scour control of major structures such as water taking, water distribution, water control, water releasing and water escape, and propose primary analysis results. Describe the calculation conditions and methods of scouring and deposition of crossing structures of important rivers, and propose primary analysis and calculation results.

6.11.3 Propose calculation conditions and methods of stability, stresses, deformation of such structures as head works, main canal system, storage and regulation works, and propose the preliminary calculation results. Study and justify the deep-buried long tunnel, large span elevated aqueduct and high head water conveyance structures if necessary.

6.11.4 Propose foundation treatment measures of such structures as head works, main canal system and storage, and regulation works.

6.11.5 Propose engineering measures for flood control, sediment scouring, seepage control and freezing resistance.

6.11.6 Based on the typical field plan, preliminarily determine the layout of farmland irrigation and drainage canal system in the typical field. Calculate the quantities of farmland works in the typical area. Estimate the quantities of farmland works in the irrigation area.

6.12 Water Supply Works

6.12.1 Basically select the water source, mode of water conveyance, and the layout of water storage and regulation works and water conveyance works.

6.12.2 Basically select the structural type, control elevations and main dimensions of structures of water intaking, water conveyance, water storage and regulation, controlling, crossing and appurtenant structures of pipes. Summarize the number of structures and main characteristic indices. Study and justify the deep-buried long tunnel, large span elevated aqueduct and high head water conveyance structures if necessary.

6.12.3 Describe computation conditions and methods of flow rate of water conveyance canal system, and hydraulic calculation conditions and methods of water levels, hydraulic gradient line, etc., and propose primary calculation results. Describe the calculation conditions and methods of the discharge capacity, hydraulic connection, energy dissipation and scour control of major structures such as water intaking, water distribution, water control, water releasing and water escape, and propose primary calculation results. Describe the calculation conditions and methods of sediment scouring and deposition of the crossing structures of important rivers and canals, and propose primary analysis results.

6.12.4 Describe calculation conditions and methods of stability, stresses, deformation of such structures as water intaking, water conveyance, storage and regulation, controlling, crossing and main appurtenant structures, and propose primary analysis results.

6.12.5 Propose foundation treatment measures of main structures.

6.12.6 Propose engineering measures for sediment scouring, seepage control and freezing resistance.

6.13 Navigation and Fish Passing Structures

6.13.1 Conduct design of navigation structures from following aspects:

  1 Describe the grade, standard, water level, type and line number of channels. Basically select the type, site and layout of the navigation works.

  2 According to the design scale, design water level, line number and lift number of navigation lock, select the layout of navigation lock, the layout plan and control sizes of lock chambers, upstream and downstream lock heads and approach channels. Preliminarily select the layout of the entrance area and connection section, the layout and structural type of navigation and berthing structures, as well as revetment and apron of the structures.

3 Describe calculation conditions and methods of stability, stresses, deformation, and permeation stability of the navigation lock structures, and propose preliminary calculation results.

4 Preliminarily select the type and layout of ship lift, lifting type, components composition and the layout and control parameters of such appurtenant structures as power supply system, lifting system, etc., propose preliminary calculation results of stability, stresses and deformation.

5 Preliminarily select foundation treatment measures of navigation structures.

6.13.2 Conduct design of fish passing structures from following aspects:

1 Basically select the type and location of fish passing structures, and propose the layout of structures of fish pass and fish lift.

2 In combination with the preliminary hydraulic calculation results of fish passing structures and their location in the project, preliminarily select major design parameters, structural type and controlling sizes of the fish pass and fish lift.

3 Conduct special study on the type and location of important fish passing structures.

6.14 Levee and River Training Works

6.14.1 Conduct levee design from following aspects:

1 Basically select the structural type of levee, and the layout and structural type of levee crossing structures. Summarize numbers and characteristic indices of levee crossing structures. Propose preliminary calculation results of the stability against sliding, stability against overturning, stresses, settlement deformation, seepage and permeation stability.

2 Preliminarily select the filling materials, compaction standard and design section of the levee.

3 Preliminarily select foundation treatment measures of the levee.

4 Preliminarily select protection scope and type of bank revetment works.

6.14.2 Conduct design of river training works from following aspects:

1 Preliminarily select the structural layout, structural type, control elevations and main dimensions of river training works for river controlling and bank revetment.

2 Propose hydraulic calculation of river training works, provide calculation methods and preliminary results of stability, stresses, deformation and seepage of river training works.

3 Preliminarily select foundation treatment measures of structures.

6.15　Access Works

6.15.1 Basically determine the layout of permanent roads and the grade of access roads in the project area. Preliminarily select the route design, and the type and standard of subgrade and pavement. Preliminarily select the position, type, span and length of bridge (culvert), and its load standard and foundation type.

6.15.2 For bridges (culverts) built for the water conveyance works which have to cross all kinds of roads, conduct the design from following aspects:

1 Determine the number, position, scale, grade and design standard of bridges (culverts).

2 Basically determine the structural type, length, span of the bridges, width of deck and load standard. Preliminarily select the foundation treatment scheme.

3 Basically determine the structural type and cross-section size of culverts. Preliminarily select the foundation treatment scheme.

6.16 Engineering Safety Monitoring

6.16.1 Preliminarily select general design scheme of safety monitoring system. Describe layout principle and scope of monitoring, and layout pattern of the monitoring items.

6.16.2 Preliminarily select the arrangement and type selection of instruments at the monitoring sections and points of different monitoring items.

6.16.3 Propose the requirements for automation of safety monitoring.

6.17 Reinforcement of Existing Projects

6.17.1 Describe the main conclusions of safety assessment of the reinforcement works, and main detection conclusions of the existing structures. For the levee to be reinforced, describe the danger over the years, and provide stability and seepage calculation results.

6.17.2 Based on necessary check and schemes comparison, preliminarily select the design scheme of reinforcement measures.

6.17.3 Preliminarily select treatment measures for the foundation or surrounding rock and connection of new and old structures.

6.18　Figures, Tables and Appendices

6.18.1　The following figures may be attached to this chapter:

1　Layout of the project sites (alignment) schemes comparison.

2　General layout of the recommended project site (alignment).

3　Layout and cross-section of the alternatives of the main structures.

4　Layout and cross-section of main structures of the recommended scheme.

6.18.2　The following tables may be attached to this chapter:

1　Schemes comparison for selection of project site (alignment).

2　Schemes comparison for dam types selection.

3　Schemes comparison for type of major structures.

4　Schemes comparison for general layout of the project.

5　Features of longitudinal and transverse profile for long route works, and quantity statistics of the structures.

6　Summary sheet for bill of quantities.

7　Others.

6.18.3　The following appendices may be attached to this chapter:

1　Special report on the calculation of important structures.

2　Report on model tests of river works, hydraulic structure and sediment.

3　Special report of new technology, new materials and new

type of structures.

4 Negotiation agreements on scheme of crossing works with the administrative authorities of highway and railway.

5 Other special reports.

7 Electromechanical Equipment and Hydraulic Steel Works

7.1 Hydraulic Machinery

7.1.1 Select turbines and auxiliary equipment from following aspects:

1 Through schemes comparison and justification, basically select basic parameters of turbines such as type, number and unit capacity. Preliminarily select the model and setting elevation.

2 Basically select the type, number and layout of the auxiliary equipment of turbine.

3 Propose calculation results of transient process (regulation guarantee) for unit.

7.1.2 Select pumps and their auxiliary equipment from following aspects:

1 Through comparison and justification of various schemes, basically select such parameters as characteristic lift, pump type, number of units, flow rate of a unit and motor power. Preliminarily select pump model and determine the setting elevation.

2 Basically select the type, number and layout of the auxiliary equipment of pumps.

3 Basically select the type of inlet and outlet conduits and shutoff means. Preliminarily select dimensions and profile of the conduit.

4 Basically select the operation mode of pump unit.

7.1.3 Basically select the type, number and layout of hoisting equipment and auxiliary equipment in the power plant.

7.1.4 For power station and pump station with heavy sediment

- laden flow passing through, propose anti-abrasion measures for main water passage components.

7.1.5 For pressure water conveyance system, preliminarily select specifications, numbers and layouts of various valves (such as regulating valve along the pipeline), flow meter, and other pipeline accessories.

7.2 Electrical System

7.2.1 Basically determine the voltage level and the number of circuits of the outgoing transmission line, interconnection points to the power system and the distance of connection.

7.2.2 Basically determine the voltage level of connection to power system of the pump (sluice) station, and the number of circuits of power supply line and connection points to the power system and the distance of connection.

7.2.3 Propose the statistical results of electric loads, basically determine the load level, and basically define the mode of power supply.

7.2.4 Basically select the main electrical connection scheme and the connection mode of the powerhouse (pump station) service power.

7.2.5 Basically select types, specifications, quantities and main technical parameters of main electrical equipment including generators, motors, main transformers, high-voltage distribution devices, generator circuit breakers, high-voltage cables, busbar, etc.

7.2.6 Basically select the layout of main electrical equipment such as main transformers and high-voltage distribution devices.

7.2.7 Preliminarily propose the scheme of over-voltage protection, and preliminarily select the earthing design scheme of the whole powerhouse (pump station).

7.2.8 Basically select the design scheme and the configuration of main equipment of monitoring, relay protection, excitation, operation and control power supply.

7.2.9 Basically select the design scheme of communication, preliminarily select the configuration of main equipment.

7.2.10 Basically select the location and layout of the monitoring center.

7.2.11 Conduct special studies for the design of SCADA system and communication system of particular multi-purpose projects and water conveyance projects. Basically select the design scheme, the configuration and layout of main equipment.

7.3 Hydraulic Steel Works

7.3.1 Basically select the type, number, size, main technical parameters and layout of gates, trash racks, valves and hoisting equipment for various kinds of hydraulic structures.

7.3.2 Basically select the type, number, size, main parameters and layout of hydraulic steel works and the hoisting equipment of the navigation and fish passing structures.

7.3.3 Describe operation principles, and propose preliminary conditions in manufacturing, transportation, installation and maintenance. Preliminarily select measures and design plan for prevention from corrosion, freezing, clogging, cavitation, abrasion and vibration.

7.3.4 Estimate the quantities of hydraulic steel works

7.4 Heating, Ventilation and Air Conditioning (HVAC)

7.4.1 Basically select HVAC design scheme, the type, number and layout of main equipment.

7.4.2 For the powerhouse with type of inside-dam, semi-underground or underground, propose the preliminary calculation results of HVAC and emergency smoke exhaust.

7.5 Fire Fighting

7.5.1 Preliminarily select the general design scheme of fire fighting for the powerhouse (pump station), powerhouse (pump station) area, main structures, navigation facilities, etc.

7.5.2 Preliminarily select the fire-fighting design of structures and major electromechanical equipment.

7.5.3 Preliminarily select the type and layout of main fire-fighting equipment.

7.5.4 Make a list of main fire-fighting equipment.

7.6 Figures and Tables

7.6.1 The following figures may be attached to this chapter:

1 Synthetic characteristic curve of turbine (pump) hydraulic model.

2 Synthetic characteristic curves of various operating conditions for single or multi-pumps in parallel operating at constant or variable speed.

3 Schematic diagram of oil, compressed air, water supply and drainage, hydraulic monitoring systems of the power plant (pump station).

4 Geographical wiring diagram of connection to power system of the power plant (pump station).

5 Main electrical connections.

6 Layout of switchyard (substation).

7 The structure diagram of SCADA system and communication system.

      8 Configuration diagram of relay protection.
      9 General layout of hydraulic steel works.
      10 General of the gate leaf and slot for particular gate.

7.6.2 The following tables may be attached to this chapter:
      1 Summary of main hydraulic machinery equipment.
      2 Summary of main electrical equipment.
      3 Summary of quantities of hydraulic steel works.
      4 Main equipment of HVAC.

8 Construction Planning

8.1 Construction Conditions

8.1.1 Brief basic conditions of project layout, conditions of construction sites, hydrology, meteorology, sediment, ice regime, geology, etc.

8.1.2 Brief the condition of external access (railway, highway and waterway) and transportation.

8.1.3 Brief the sources of construction materials, the supply conditions of water, power and communication, and the possible capability of local maintenance and machinery repair.

8.1.4 Brief the requirements on water drainage, ice releasing, water supply and navigation during construction (including initial reservoir impoundment period).

8.1.5 Describe the requirements and comments of relevant authorities on the project in construction duration.

8.2 Selection and Exploitation of Borrow Area/Quarry

8.2.1 Analyze the distribution, reserves, quality, exploitation and transport conditions of the sand and gravel, rock, and soil in various borrow areas/quarry sites, as well as exploitation yield, utilization ratio and other technical parameters. Through technical and economic comparison, basically select the borrow/quarry areas, propose the exploitation techniques, and select the exploitation, transportation and processing equipment.

8.2.2 Determine the utilization mode and ratio of various excavated materials.

8.2.3 If necessary, determine the way of material supply after technical and economic comparisons of separate borrow area/quarry and materials purchased from the market.

8.3 River Diversion and Closure During Construction

8.3.1 Determine the grade of river diversion works. Select flood standard of river diversion and temporary water retaining of the dam in various construction duration. Determine the construction duration, and discharge for diversion works. Select the river diversion mode.

8.3.2 Basically select the layout, structural type and foundation seepage control measures for the river diversion works at each stage. Propose corresponding quantities of the projects. Propose main results of hydraulic calculations. Carry out the diversion model test if necessary. Preliminarily select measures for ice releasing and navigation during construction.

8.3.3 Basically select the mode, period and discharge of river closure.

8.3.4 Basically determine the combination mode of river diversion/closure structures and permanent structures.

8.3.5 Basically select the river closure period, gate closing discharge and scheme of closure. Propose downstream temporary water supply measures.

8.3.6 Basically select construction methods and procedures for the river diversion and closure works. Elementarily select main machineries and equipment.

8.4 Construction of Main Works

8.4.1 Basically select construction methods and procedures of main works. Describe the construction procedure, methods, arrangement and schedule for earth – rock excavation. Determine the scheme of utilization, spoiling and transportation of excavated materials. Describe measures of material preparation, transportation, filling and compaction for earth – rock filling. Describe the construction procedure, methods and arrangement for the concrete placement of each construction duration. In addition, for the underground works, describe the excavation scheme, supporting and safety protection measures.

8.4.2 Basically select main construction machineries and equipment required for the main works.

8.4.3 Conduct special study on the construction schemes if adopting new technology, new process, new equipment and new materials, or with difficult construction conditions and less construction experience.

8.5 Construction Transportation and General Construction Layout

8.5.1 Select the scheme of external access and transportation. Determine the standard and layout of main on – site trunk roads. Propose corresponding quantities.

8.5.2 Basically determine the scale and layout of main construction plants and living facilities of aggregate processing plant, concrete batching plant, concrete pre – cooling (or preheating) system, mechanical repair, general processing system, air, water and power supply and communication system.

8.5.3 Analyze and calculate the excavation – fill balance of earth – rock work. Basically select the layout of deposit and spoil areas. The planning and layout shall be carried out in section for the route works.

8.5.4 Basically select general construction layout. Propose quantities of temporary works, area and scope of construction land use. Propose the special report of general construction planning if necessary.

8.6 General Construction Schedule

8.6.1 Describe the principles and basis for preparation of the general construction schedule.

8.6.2 Propose critical schedule of pre-preparation, preparation, construction of main works and project completion period. Describe the critical schedule and corresponding construction intensity at each stage. If necessary, describe the schedule of river diversion and closure, foundation pit dewatering, flood retaining, river diversion at later stage, closing gate for initial reservoir impoundment, etc.

8.6.3 Propose general construction schedule. Basically determine the total construction duration.

8.6.4 Describe major quantities and the proportion to be completed by year. Propose the quantities of materials and labor force demanded by the project.

8.7 Figures, Tables and Appendices

8.7.1 The following figures may be attached to this chapter:

    1 Sketch of external access and transportation.

    2 Layout and structure of river diversion and closure works.

    3 Construction procedures, construction methods and construction layout of main structures.

    4 General construction layout.

    5 Scope of construction site.

    6 General construction schedule.

8.7.2 The following tables may be attached to this chapter:

    1 Summary of quantities.

    2 Main construction equipment.

8.7.3 The following appendices may be attached to this chapter:

    1 Special reports on general construction plan.

    2 Other related appendices.

9 Land Acquisition and Resettlement

9.1 General Description

9.1.1 Brief natural and socioeconomic conditions of the region involved in land acquisition.

9.1.2 Brief principal achievements and status of the review and approval on the preliminary land acquisition and resettlement plan at the project proposal stage.

9.1.3 Brief review and approval status of the plan outline of land acquisition and resettlement.

9.1.4 Brief principal results of land acquisition and resettlement plan at this stage.

9.2 Scope of Land Acquisition

9.2.1 Determine the design flood standards for the areas of reservoir inundation and locations of residents to be resettled, industrial enterprises and other specific works. Determine the principles of land acquisition for the construction of the main structures and other relative works.

9.2.2 Based on results of reservoir backwater computation and engineering geological investigation in the project area, determine the range of reservoir – affected area. Based on the design of general project layout, construction planning and project management, determine the range of land acquisition for the construction of the main structures and other relative works.

9.3 Land Acquisition Objects

9.3.1 Identify objects in the land acquisition area, and prepare the inventory investigation report of land acquisition.

9.3.2 Investigate socioeconomic condition of the reservoir area, the construction area of multi-purpose project and other water conservancy works and resettlement area. Evaluate the impact of land acquisition on local economy and society.

9.3.3 Describe organization, time, contents and methods of the inventory investigation of land acquisition.

9.3.4 Describe inventory investigation results in rural area.

9.3.5 Describe fundamental situation of towns (fairs) involved in land acquisition, socioeconomic conditions in the region, inventory investigation results and the impact of land acquisition on the towns (fairs).

9.3.6 Describe fundamental situation of industrial enterprises involved in land acquisition, inventory investigation results, and impact of land acquisition on the industrial enterprises.

9.3.7 Describe quantity, grade, scale and impact of the facilities of transportation, power transmission and substation, telecommunications, broadcast/television, the water and hydropower projects, and the objects of mine resources, historic relics, etc. to be involved in land acquisition of project construction.

9.3.8 Describe inventory investigation results involved in land acquisition of the schemes of the project.

9.3.9 Explain reasons for the change when the inventory investigation results are different from that at the project proposal stage.

9.4 Rural Resettlement

9.4.1 Determine the design reference year of resettlement plan, natural population growth rate and resettlement standard. Calculate and determine the population of employment-oriented resettlement and the population of relocation resettlement taking villager group as a statistic unit for the reservoir project, while taking the administrative village as a statistic unit for other water resources projects.

9.4.2 According to the recommended construction scheme, analyze and determine the resettlement environmental carrying capacity taking administrative village as a statistic unit for the reservoir project, while taking township (town) or administrative village as a statistic unit for other water resources projects.

9.4.3 Determine the rural resettlement locations and the employment-oriented resettlement mode by taking administrative village as a statistic unit according to the requirements of local residents (resettled residents and residents in the relocation places) and government, and taking the results of environmental capacity analysis in consideration; analyze the balance of employment-oriented resettlement plan cost, and compile the rural employment-oriented resettlement plan.

9.4.4 Based on the recommended development scheme, the requirements of local residents (resettled residents and residents in the relocation places) and government, and the employment-oriented resettlement plan, determine rural resettlement location, resettlement population, and land use area. Basically identify the engineering geological and hydrogeological conditions of the proposed concentrated residential area, and evaluate the stability of sites, adaptability of structures and the geological hazards. Choose representative residential area for topographic mapping with scale no less than $1:1000$ and carry out design of the residential area; and prepare the rural resettlement plan.

9.4.5 Propose the plan of post-resettlement support.

9.5 Town (Fair) Relocation

9.5.1 Determine the basis and principle of town (fair) relocation plan.

9.5.2 After collecting opinions of local government and resettlers, determine the mode of town (fair) relocation. Basically ascertain the engineering geological and hydrogeological conditions of the proposed resettlement area. Choose the relocation place.

9.5.3 Determine the population, available land and construction standards of infrastructures of town (fair).

9.5.4 Compile detailed controlling and construction plan of city (town).

9.6 Disposition of Industrial Enterprises

9.6.1 Determine the basis and principle for disposition plan of industrial enterprises.

9.6.2 Propose disposition schemes of industrial enterprises.

9.7 Disposition of Specific Works

9.7.1 Determine the basis and principle for disposition of specific works.

9.7.2 Determine the disposition schemes of specific works. For the specific works with great importance, large scale and high cost, carry out typical design in accordance with the requirements of primary design stage as per corresponding disciplines, and propose design documents.

9.7.3 Propose measures for protection of historic relics and treatments of covered mineral resources.

9.8 Protection Works

9.8.1 Determine the design basis, principle and standard of protection works.

9.8.2 For important submerged objects with condition for protection, justify and determine protection scheme, and propose design documents.

9.9 Clearance of Reservoir Basin

9.9.1 Determine the range and content of reservoir basin clearance.

9.9.2 Identify the type and scale of objects to be cleared, and clearance quantity.

9.9.3 Propose technical requirements and measures of reservoir basin clearance.

9.9.4 Prepare the reservoir basin clearance plan.

9.10 Figures, Tables and Appendices

9.10.1 The following figures may be attached to this chapter:
    1 Diagrammatic map of scope of land acquisition.
    2 Diagrammatic map of resettlement plan.
    3 Others.

9.10.2 The following tables may be attached to this chapter:
    1 Inventory investigation results.
    2 Calculation of employment – oriented resettlement population.
    3 Calculation of relocated resettlement population.
    4 Employment – oriented resettlement plan.
    5 Relocation resettlement plan.
    6 Cost estimate of compensation for land acquisition and

resettlement.

7 Others.

9.10.3 The following appendices may be attached to this chapter:

1 Report on inventory investigation of land acquisition (including notice issued by provincial government to forbid constructing new project and new resettlement in the land acquisition area and the comments of local government on inventory investigation results of land acquisition).

2 Comments of local government on resettlement plan.

3 Approval comments on resettlement plan outline.

4 Design report on relevant individual project.

5 Comments of relevant departments and units on treatment of industrial enterprises and special projects.

6 Related documents of agreements, contracts, commitments, etc.

7 Others.

10 Environmental Impact Assessment

10.1 General Description

10.1.1 Brief preparation conditions of environmental impact report (tables) and main conclusions at the project proposal stage.
10.1.2 Describe the standard and basis for environmental impact assessment.

10.2 Investigation and Assessment of Environmental Status

10.2.1 Brief the regional environmental status and main environmental issues.
10.2.2 Investigate and assess the environmental status in the project affected area. Define the main environmental sensitive points. Analyze the main environmental issues and variation tendency.
10.2.3 Determine the objectives of environmental protection.

10.3 Environmental Impact Prediction and Assessment

10.3.1 Analyze hydrological regime changes at the main control section in the different typical years before and after the project operation. Evaluate the satisfaction of water use for river ecological environment.
10.3.2 For urban water supply project, evaluate the status of water quality, and predict the water quality at the design reference year.
10.3.3 Predict and assess the ecological impacts from following

aspects:

1 Prediction and assessment on terrestrial ecological impact.

2 Prediction and assessment on aquatic ecological impact.

3 Impact and assessment on sensitive objectives of ecological protection.

10.3.4 Predict and assess the water environmental impacts from following aspects:

1 Impacts of the eutrophication in the reservoir area and the water environment in the flow reducing river reach.

2 Impact of water evacuation due to irrigation on water environment.

3 Impact of flow reducing on seawater invasion in the estuary area.

4 Impact of water temperature variation on irrigation and downstream aquatic ecology.

10.3.5 Predict the impacts of soil environment and land resources from following aspects:

1 Variation of regional groundwater level after project operation.

2 Improvement of soil environment due to project construction, and the location and extent of soil gleization, soil paludification, soil secondary salinization and soil pollution.

3 Impacts of reservoir inundation and project construction on land resources.

10.3.6 For public health impact prediction, analyze the impact of spreading diseases, such as natural focal diseases, water-borne epidemic diseases, insect-borne infectious diseases and endemic diseases caused by project construction to the health of construction personnel and local residents.

10.3.7 For resettlement impact prediction, analyze the impact of resettlement on ecology, water environment and public health.

10.3.8 For environmental impact during the construction duration, predict the impacts of construction waste water, waste gas, noise and solid waste on sensitive objects such as receiving water body, atmospheric environment and noise.

10.3.9 For other environmental impacts, predict and assess the impacts on sediment, local climate, environmental geology, landscape, heritage, religion, culture, society and economy according to the project and environmental characteristics.

10.3.10 Comprehensively assess the environmental rationality of the project design scheme. Propose the environmental recommendation comments and protection requirements.

10.4 Environmental Protection Countermeasures

10.4.1 Basically determine protection measures to the water source of the urban water supply project

10.4.2 Basically determine aquatic environmental protection measures to the affected water intakes and groundwater users. Basically determine measures to mitigate the effect of low temperature water.

10.4.3 Basically determine protection countermeasures of rare and endangered fauna and flora, and aquatic species.

10.4.4 Basically determine countermeasures for soil quality protection and pollution prevention.

10.4.5 Basically determine countermeasures for public health protection.

10.4.6 Basically determine pollution prevention and control measures during construction.

10.4.7 Propose other environmental protection countermeasures.

10.5　Environmental Management and Monitoring

10.5.1 Propose the environmental management program during construction and operation periods.

10.5.2 Propose the environmental monitoring plan during construction and operation periods.

10.6　Comprehensive Assessment Conclusion

10.6.1 Brief environmental impact assessment conclusions to main environmental factors.

10.6.2 Brief main environmental protection countermeasures.

10.6.3 Propose comprehensive conclusion of environmental impact assessment.

10.7　Figures and Appendices

10.7.1 The following figures may be attached to this chapter:

1 Distribution of main environmental protection objects.

2 Function zoning of important ecological and environmental protection objects.

3 General layout of environmental protection measures.

4 Design of important ecological protection measures.

5 Layout of environmental monitoring station/network.

10.7.2 The following appendices may be attached to this chapter:

1 Comments on the environmental impact assessment at planning stage.

2 Comments of related authorities.

11 Water and Soil Conservation

11.1 General Description

11.1.1 Brief the main conclusions of the project at the technical proposal stage and preparation of the water and soil conservation report.

11.1.2 Describe the natural conditions, soil erosion and water and soil conservation in the project area.

11.1.3 Describe the conditions and relevant requirements of national or provincial key soil erosion areas and key management areas in the project area.

11.2 Water and Soil Conservation Assessment for Main Works

11.2.1 Make water and soil conservation assessment on the general layout of main works, planning of disposal areas, construction planning, etc. Define whether there are some restrictive problems on water and soil conservation, and propose the solution suggestions.

11.2.2 Propose the water and soil conservation requirements and suggestions on the design of main works according to the conclusions of assessment.

11.3 Responsible Area and Zoning of Soil Erosion Control

11.3.1 Define the principles and methods of determining the responsible area of soil erosion control.

11.3.2 Determine the area and distribution of responsible area of soil erosion control. Describe the relationship between the responsibility scope of soil erosion and land acquisition.

11.3.3 Determine the zoning of water and soil erosion control.

11.4 Prediction of Soil Erosion

11.4.1 Determine disturbed land area due to project construction, quantity of disposal soil, rock and residue, as well as types and number of damaged water and soil conservation facilities.

11.4.2 Basically determine the prediction methods, period and main parameters of soil erosion. Predict potential soil erosion area and increased quantity of soil erosion due to project construction. Analyze the possible hazards.

11.5 Standard and General Layout of Soil Erosion Control

11.5.1 Determine the standard grade and target of soil erosion control.

11.5.2 Determine the system and general layout of soil erosion control measures. In accordance with the selected disposal area, volume of waste and topographical conditions, basically determine the type of disposal area, stacking plan, general arrangement of protective measures, as well as control measures of other zonings.

11.6 Design of Zoning Control Measures

11.6.1 Determine the level and design standard of water and soil conservation.

11.6.2 Make typical design on all kinds of control measures, and propose the quantities of water and soil conservation.

11.6.3 Propose the construction planning of water and soil conservation.

11.7 Water and Soil Conservation Management and Monitoring

11.7.1 Propose monitoring plan of water and soil conservation.

11.7.2 Propose water and soil conservation management requirements or program during construction and operation periods.

11.8 Figures and Appendices

11.8.1 The following figures may be attached to this chapter:

　　1 General layout of responsible area and measures of soil erosion control.

　　2 Typical design of water and soil conservation.

　　3 Construction schedule of water and soil conservation.

　　4 Layout of water and soil conservation monitoring points.

11.8.2 Other necessary appendices may also be attached to this chapter.

12 Labor Safety and Industrial Hygiene

12.1 Analysis of Dangerous and Harmful Factors

12.1.1 Describe the design basis of laws and regulations, principal technical standards and relevant documents.

12.1.2 Brief the overview of project design.

12.1.3 According to the natural and social conditions and surrounding situation of project site, basically determine the dangerous and harmful factors which affect labor safety and industrial hygiene during project construction and operation as well as their harmful degree.

12.1.4 Basically determine the harmful factors and extent to labor safety and industrial hygiene considered in selection and arrangement of hydraulic structures and electromechanical equipment.

12.1.5 Basically determine the harmful factors and extent to labor safety and industrial hygiene in the temporary construction facilities.

12.2 Labor Safety Measures

12.2.1 Propose the requirements and design principles for preventing mechanical injuries, electrical injuries, falling injuries, airflow injuries and injuries due to strong wind, fog, rain, thunder and lightning. Basically determine protective measures.

12.2.2 Propose the requirements and design principles for preventing injuries from landside and debris flow, flood, fire and explosion and access accidents. Basically determine protective measures.

12.2.3 According to factors impacting labor safety, analyze and propose other principles and measures for safety design.

12.3 Industrial Hygiene Measures

12.3.1 Propose the requirements and design principles for preventing impact of harmful factors such as noise and vibration, electromagnetic radiation, dust and pollutant, leakage of radioactive material and poison, etc. Basically determine protection measures or measures for mitigating and avoiding impact.

12.3.2 Propose the requirements and design principles for daylight and illumination, ventilation, temperature and humidity control, waterproof and moisture – proof in various workplaces. Basically determine industrial hygiene guarantee measures.

12.3.3 Basically determine hydraulic schistosomiasis control measures for the project in schistosomiasis disease area.

12.3.4 Describe the requirements of safety and hygiene management and facilities configuration for the rooms of production, test, office and residence among permanent buildings. Propose organizations and facilities configuration for safety and hygiene management.

12.3.5 According to factors impacting industrial hygiene, analyze and propose other design principles of industrial hygiene.

12.4 Safety and Health Assessment

12.4.1 Analyze and assess various protective measures for labor safety.

12.4.2 Analyze and assess the design of facilities configuration for industrial hygiene.

13 Energy Saving Evaluation

13.1 Design Basis

13.1.1 Brief the reasonable energy use standard and energy saving design specifications for the project to be abided by.

13.1.2 Brief energy supplies, energy consumption and main indices of the place where the project is located. Define middle and long term special plan and energy conservation goals developed by national and local governments as well as the industry.

13.2 Energy Consumption Analysis

13.2.1 Analyze the energy demand and supply of the place where the project is located.

13.2.2 According to specific circumstances of the project construction, estimate the total energy consumption and varieties during construction and operating periods of the project.

13.2.3 In accordance with the requirements of national and local energy conservation targets and specific conditions of the project construction, basically determine the energy consumption indices during the construction and operation periods of the project.

13.3 Energy Conservation Measures

13.3.1 Propose general layout of the project and principles and requirements of energy conservation design related to structures, construction planning, electromechanical equipment and related management facilities.

13.3.2 Basically determine energy conservation measures of the project.

13.3.3 Analyze total energy consumption during construction and operation periods.

13.4 Energy Conservation Effect Evaluation

13.4.1 Analyze whether the project is in compliance with national, local, and industrial design requirements for energy conservation.

13.4.2 Evaluate the feasibility of energy conservation measures on general layout, construction planning, electromechanical equipment and related management facilities.

13.4.3 Evaluate the energy conservation effect.

14　Project Management

14.1　Project Management System

14.1.1　Describe relevant stipulations of the state and comments on approval of the project proposal. Basically determine the type and nature of the management agency.

14.1.2　Basically determine the management system in project operation period, and describe the administrative relation and ownership of assets.

14.1.3　Propose the institutional settings and staffing of management agency in the operation period, and describe its duty.

14.1.4　Propose the institutional settings of management agency in the construction duration, and propose the tendering/bidding scheme for project construction.

14.2　Project Operation Management

14.2.1　Describe the management requirements and management contents of routine maintenance, safety monitoring, regulation and operation of the project. For the project consisting of multiple single projects, define the management relationship of each single project.

14.2.2　Brief regulation principles and modes of the project.

14.2.3　Describe the operational cost and source for maintaining project operation, and propose corresponding policy recommendations if necessary.

14.2.4　Attach commitment letter or letter of intent of relevant authorities on underwriting of water and electricity.

14.3 Scope of Management and Protection

14.3.1 Determine the scope of management and protection of the project.

14.3.2 Determine the area of land use in the scope of management.

14.3.3 Propose the management and restriction requirements on land use in the scope of management and protection.

14.4 Management Facilities and Equipment

14.4.1 Basically determine the area, land use and location of management region required for the project management agency.

14.4.2 Basically determine contents and quantity of facilities for production, cultural welfare, transportation and communication, etc. For the renovation, extension and reinforcement works, describe existing facilities and service conditions of the management agency.

14.4.3 Describe the infrastructures construction of engineering observation, monitoring, communication and dispatching system and hydrological telemetry system.

15 Cost Estimate

15.1 General Description

15.1.1 Brief the project conditions, and describe indices such as project scale, purpose, main quantities, quantity of main materials, total construction duration, land occupation, inundated land and relocated population.

15.1.2 Describe main indices of cost estimate.

15.2 Principle and Contents of Preparation

15.2.1 Prepare cost estimate for the engineering part of the project from following aspects:

1 Describe specifications, quotas and other relevant provisions, price reference year for cost estimate, and basis for pricing of main materials, subsidiary materials, electromechanical equipment, hydraulic steel works and aggregates. Describe the time, document number and applicable conditions of relevant provisions and quotas issued by other industries.

2 Define the itemization of cost estimate according to Sinohydro [2002] No.116 *Provisions for preparation of estimates for the design of water conservancy project*, and classification of the project.

3 Analyze and calculate budgetary prices of main materials. Determine prices of subsidiary materials. Calculate basic unit prices and unit prices of works according to the construction planning.

4 Investigate, analyze and determine cost indices of transportation, buildings, transmission lines and other works.

5 Investigate, analyze and determine prices of main electromechanical equipment and hydraulic steel works.

6 For the project with foreign investment, describe the form and basis for using foreign investment, and calculate the cost based on all domestic cost estimates and thereafter consider cost implication of using foreign investment.

15.2.2 Prepare cost estimate for the compensation of land acquisition and resettlement from following aspects:

1 Describe specifications, quotas and other relevant provisions, and price level year for cost estimate.

2 Analyze and determine the compensation and subsidy standards for various types of land acquisition. Determine compensation unit prices of houses and appurtenances.

3 Determine unit prices and cost of main items, such as rural residential areas, towns (fairs), specific items, industrial and mining enterprises, protection works and clearance of reservoir bottom.

4 Determine other fees and costs.

5 List relevant taxes, duties and levies according to relevant regulations.

15.2.3 Prepare cost estimate for environmental protection from following aspects:

1 Describe specifications and reference documents for cost estimate of environmental protection.

2 Estimate cost of environmental protection measures, environment monitoring measures, instrument, equipment and installation, and temporary environmental protection measures separately.

15.2.4 Prepare cost estimate for water and soil conservation from following aspects:

1 Describe specifications, quotas and that of other relevant industries for cost estimate of water and soil conservation.

2 Analyze and calculate main base unit prices and engineering unit prices according to the price level in the cost estimate preparation year.

15.3 Results of Cost Estimate

15.3.1 Results of cost estimate shall contain the cost estimate report (original) and appendices.

15.3.2 The cost estimate report (original) shall contain the followings:

1 Description on the preparation, including the project conditions, principles and basis of preparation and main cost indices, etc.

2 Summary table of cost estimate (including cost of the engineering part, land acquisition and resettlement compensation, environmental protection and water and soil conservation).

3 Cost estimate table of the engineering part shall contain the followings:

    1) Total cost estimate.
    2) Estimate of civil works.
    3) Estimate of electromechanical equipment and installation works.
    4) Estimate of hydraulic steel works and installation works.
    5) Estimate of temporary construction works.
    6) Estimate of independent costs.
    7) Annual cost table.
    8) Capital flow statement.
    9) Summary of unit prices of construction works.
    10) Summary of unit prices of installation works.
    11) Summary of budgetary prices of main materials.

12) Summary of budgetary prices of subsidiary materials.

13) Summary of machine-hour rates of construction machineries.

14) Summary of quantities.

15) Summary of quantities of main materials.

16) Summary of quantities of man-hours.

4 Cost estimate table of land acquisition and resettlement compensation may contain the followings:

1) Total cost estimate.

2) Analysis of unit prices of compensation items.

3) Breakdown of compensation cost.

4) Annual cost table.

5 Cost estimate table of environmental protection may contain the followings:

1) Total cost estimate.

2) Estimate of environmental protection measures.

3) Estimate of environment monitoring measures.

4) Estimate of environmental protection instrument and installations.

5) Estimate of temporary works.

6) Estimate of independent costs.

7) Annual cost table.

8) Estimate of other individual works.

6 Cost estimate table of water and soil conversation may contain the followings:

1) Total cost estimate.

2) Estimate of engineering measures.

3) Estimate of vegetation measures.

4) Estimate of temporary works.

5) Estimate of independent costs.

6) Annual cost table.
7) Summary of unit price of main works.
8) Summary of unit price of main materials.
9) Summary of machine – hour rates of construction machineries.

15.3.3 Appendix of cost estimate shall contain the followings:

　1　Cost estimate table of the engineering part shall contain the followings:

1) Unit price of labor budget.
2) Transportation costs of main materials.
3) Budgetary prices of main materials.
4) Calculation of construction power prices.
5) Calculation of construction water prices.
6) Calculation of construction compressed air prices.
7) Calculation of supplementary quotas.
8) Calculation of supplementary machine – hour rates of construction machineries.
9) Calculation of unit prices of aggregates.
10) Unit prices of concrete materials.
11) Unit prices of construction works.
12) Unit prices of installation works.
13) Calculation of freight and miscellaneous expenses of main equipment.
14) Calculation of the cost of temporary housing construction works.
15) Calculation of independent costs.
16) Capital flow statement.
17) Price contingency.
18) Calculation of financing interest during construction duration.

2 Cost estimate table of environmental protection may contain the followings:

1) Calculation of unit price of labor budget.
2) Calculation of transportation costs of main materials.
3) Calculation of budgetary prices of main materials.
4) Unit prices of construction works.
5) Unit prices of installation works.
6) Calculation of independent costs.

3 Cost estimate table of water and soil conversation may contain the followings:

1) Unit price of labor budget.
2) Transportation costs of main materials (including sapling and seeds).
3) Budgetary prices of main materials.
4) Calculation of prices of construction power, water and compressed air (attach calculation explanation for any difference from main works).
5) Calculation of machine-hour rates of construction machineries (attach calculation explanation for any difference from main works).
6) Calculation of unit prices of aggregates (including sapling and seeds) (attach calculation explanation for any difference from main works).
7) Unit prices of concrete materials (attach this table for any difference from main works).
8) Unit prices of engineering measures.
9) Unit prices of vegetation measures.
10) Unit prices of temporary works.
11) Calculation of independent costs (attach calculation explanation).

15.4 Cost Comparison and Analysis

15.4.1 Describe cost changes of engineering part, land acquisition and resettlement compensation part, environmental protection part and water and soil conversation part from those at project proposal stage. Analyze the causes in aspects of price fluctuation, items and quantities adjustment, national policy change, etc., and explain the analysis conclusions.

15.4.2 Analysis on cost changes shall contain the following tables:

1 Comparison of total costs (between current stage and project proposal stage).

2 Comparison of quantities of main works.

3 Comparison of base unit prices, main materials and equipment prices (compensation unit prices).

16 Economic Evaluation

16.1 General Description

16.1.1 Brief the background, development tasks, scale, benefits, construction contents, duration, characteristics, management agency, etc. of the project.

16.1.2 Brief the basic references and calculation principles of economic evaluation.

16.2 Expenditure Estimation

16.2.1 Brief the main references, price datum year, and annual investment plan for the construction project cost (excluding interests in the construction duration). Describe the change of expenditure estimation results between current stage and project proposal stage.

16.2.2 Describe the expenditure estimation methods of cash flow, and estimate the amount of cash flow.

16.2.3 Assess the annual operation cost and total cost.

16.2.4 Describe relevant tax categories and tax rates to be paid.

16.3 National Economic Evaluation

16.3.1 Describe the references, calculation methods and the principles of parameters selection for national economic evaluation.

16.3.2 Adjust the cost of construction projects. Calculate the quantifiable external cost. Review the cost of national economic evaluation.

16.3.3 Assess the output economic benefits and the quantifiable external benefits of the proposed project.

16.3.4 Calculate the evaluation indices such as economic net present value, internal rate of return, economic cost – benefit ratio, etc.

16.3.5 Make a sensitivity analysis to the main factors impacting the national economic indices, and explain the impact of their change on national economic evaluation indices.

16.3.6 Make a comprehensive evaluation on economic rationality of the project, and review the conclusion comments.

16.4　Fund Financing Plan

16.4.1 Describe the recommended fund raising plan at the project proposal stage as well as relevant review and approval comments.

16.4.2 Describe principles and methods of multi – purpose project investment and cost allocation, review the results of cost allocation. Calculate the unit cost of water supply, irrigation and power generation.

16.4.3 Investigate supply and demand conditions of water and electricity. Investigate market prospect, and market competitiveness of the products. Propose the various schemes of water and electricity prices. Analyze the affordability of different users to water and electricity prices.

16.4.4 Describe the using conditions and calculating principles of the borrowed fund, propose the calculation program of borrowed fund. Analyze and calculate the proportion of borrowed fund to non – borrowed fund.

16.4.5 Describe sources, composition, using conditions and benefit requirements of the non – borrowed fund. Analyze the contribution proportion of each investor.

16.4.6 Make a comprehensive analysis for project financing capacity. Propose the recommended financing program. Indicate changes to the project proposal stage.

16.5 Financial Evaluation

16.5.1 Analyze the annual financial revenues, total profits, after-tax profits, statutory surplus reserve, and distributable profits for project investors.

16.5.2 Analyze the break-even balance situation and financial viability of the project.

16.5.3 Calculate the interest coverage ratio, debt service coverage ratio and assets liability ratio, and analyze the repayment ability of the project.

16.5.4 Calculate indices of the financial internal rate of return of total cost and the capital financial internal rate of return of the project. Analyze the profitability of project.

16.5.5 Analyze main factors affecting financial indices and the critical point of main sensitive factors.

16.5.6 Analyze potential economic risks of the project. Identify risk factors and analyze preliminarily the probability distribution of major risk factors if necessary.

16.5.7 Make a comprehensive evaluation for the financial feasibility of the project.

16.6 Figures and Tables

16.6.1 The following figures may be attached to this chapter:
    1 Breakeven analysis.
    2 Sensitivity analysis.
    3 Others.

16.6.2 The following tables may be attached to this chapter:
 1 Expenditure estimation of construction.
 2 Total expenditure estimation plan and financing.
 3 Total cost estimation table.
 4 Cash flow of project cost.
 5 Cash flow of capital.
 6 Financial cash flows of all investors.
 7 Profit and loss statement.
 8 Capital financial cash flows.
 9 Balance sheet.
 10 Schedule of principal and interest repayment.
 11 Debt financing capacity.
 12 Non-debt funds constitution.
 13 Economic benefit cost flow of project investment.
 14 Adjustment for investment expenditure estimation of economic analysis.
 15 Adjustment for operating expenditure estimation of economic analysis.
 16 Indirect expenditure estimation of the project.
 17 Economic benefit estimate of the project.
 18 Sensitivity analysis.
 19 Classification of risk levels.
 20 Others.

17 Risk Analysis of Social Stability

17.1 Basis of Preparation

17.1.1 Laws and regulations, policies and normative documents.

17.1.2 National guideline on regional socioeconomic development and relevant plan approved by the State Council or authorities concerned.

17.1.3 Necessary data for basic conditions of the proposed project and social stability risk analysis.

17.1.4 Relevant approval documents of the proposed project.

17.2 Risk Identification

17.2.1 The investigation of social stability risks shall consider the legal requirement, rationality, feasibility and control capability of the construction of the proposed project.

17.2.2 The investigation scope shall cover all areas and stakeholders involved in or potentially affected by the project.

17.2.3 The investigation contents shall mainly include the project plan design, social development, natural environmental conditions, sensitive objects, interest appeals, public opinions, etc.

17.2.4 The investigation shall adopt flexible methods according to local and actual situations of the project by thoroughly inquiring and soliciting views of all stakeholders.

17.3 Risk Factors Analysis

17.3.1 Identify thoroughly and comprehensively various factors that may lead to risks of social stability in respect of legality, rationality and feasibility of project construction based on the results of social stability risk investigation in thorough and comprehensive way.

17.3.2 Classify and sort order of risk factors, and define major risk factors based on the cause, probability, occurring condition and time, impact performance and scope of identified risk factors.

17.3.3 Analyze the major risk factors with qualitative and quantitative methods, and propose a preliminary conclusion of the risk analysis.

17.4 Risk Prevention and Mitigation Measures

17.4.1 Analyze controllable factors of social stability risk, and propose the prevention and mitigation measures contrapuntally based on the preliminary conclusion of the risk assessment.

17.4.2 Analyze and justify the completeness, effectiveness and feasibility of the risk prevention program and measures in the processes of project decision, construction and operation based on the analysis of controlling factors, and propose controllable opinions and recommendations.

17.4.3 Propose risk mitigation measures and recommendations to the uncontrollable factors according to the controllable opinions and recommendations, and analyze the effectiveness and feasibility of the measures.

17.5 Risk Analysis Conclusions

17.5.1 Define the main and critical social stability risk factors.
17.5.2 Propose the main social stability risk prevention and mitigation measures.
17.5.3 Propose suggestions on social stability risk level after taking risk prevention and mitigation measures.

Annex A Format of Feasibility Study Report

A. 0. 1 The report cover shall meet the following requirements:

1 The cover shall include report name, full name of the design firm, completion time (year/month), etc.

2 The report name shall include such contents as administrative region, name of the river basin where the project is located, project name and properties, etc.

3 The responsible unit shall be the first design unit when the report is accomplished by several design units.

4 The report shall be marked with "submitted for approval" "revision" and so on.

A. 0. 2 The following contents shall be included on the front pages of the report:

1 Copies of qualification and quality certificates of the design firm.

2 Signature page of the design firm. The names of approver, reviewer, chief design engineer, professional responsible persons, main drafters shall be printed. The signatures of the approver, reviewer and chief design engineer shall be affixed.

3 Project image or aerial view. There shall be pictures of the status of the extension and reconstruction projects.

A. 0. 3 List the reviewer, checker and drafter on the front page of each chapter. The list shall include such contents as professional title, serial number of registered qualification certificate and signature. Sections specified in this standard may be included or excluded in corresponding chapters of the report according to the actual situation.

A. 0. 4 When a large number of official approval and relevant documents are necessary to be included in the feasibility study report, bind them together with the executive summary of the report and print it into a single book.

A. 0. 5 Arrange the required appendices sequentially according to different disciplines and bind them separately.

Annex B Format of the Table of Project Characteristics at Feasibility Study Stage

Table B Project characteristics of ××× project at feasibility study stage

Description	Unit	Quantity	Remarks
Ⅰ. Hydrology			
1. Catchment area			
Entire catchment area	km²		
Upstream the dam (sluice) site	km²		
2. Number of hydrological series years used	a		Years with actual measurement plus interpolated and extrapolated data
3. Average annual runoff	$10^8 \mathrm{m}^3$		
4. Characteristic flow			
Average annual flow	m³/s		
Maximum observed flow	m³/s		Date of observation
Minimum observed flow	m³/s		Date of observation

Table B (Continued)

Description	Unit	Quantity	Remarks
Maximum investigated historical flow	m^3/s		Date of occurrence
Design flood discharge ($P=$ %)	m^3/s		
Check flood discharge ($P=$ %)	m^3/s		
Construction diversion discharge ($P=$ %)	m^3/s		
5. Flood volume			
Maximum observed flood volume (d)	$10^8 m^3$		Date of observation
Design flood volume (d)	$10^8 m^3$		
Check flood volume (d)	$10^8 m^3$		
6. Sediment			
Average annual suspended load discharge	$10^4 t$		
Average annual sediment concentration	kg/m^3		
Maximum observed sediment concentration	kg/m^3		Date of observation
Average annual bed load discharge	$10^4 t$		
7. Natural water level			
Average annual water level Corresponding flow	m m^3/s		Indicate the place of observation

Table B (Continued)

Description	Unit	Quantity	Remarks
Minimum observed water level	m		
Corresponding flow	m³/s		Date and place of observation
Maximum observed water level	m		
Corresponding flow	m³/s		Date and place of observation
Minimum investigated water level	m		
Corresponding flow	m³/s		Date and place of occurrence
Maximum investigated water level	m		
Corresponding flow	m³/s		Date and place of occurrence
II. Project Scale			
1. Reservoir			
Check flood level ($P=$ %)	m		
Design flood level ($P=$ %)	m		
Normal pool level (full supply level)	m		
Maximum water level for flood routing ($P=$ %)	m		
Limited water level in flood season	m		
Dead water level	m		

Table B (Continued)

Description	Unit	Quantity	Remarks
Total storage capacity (below maximum flood level)	10^8m^3		
Flood control storage (from maximum water level for flood routing to limited water level in flood season)	10^8m^3		
Active storage (from normal pool level to dead water level)	10^8m^3		
Dead storage (below dead water level)	10^8m^3		
Reservoir area at normal pool level	km^2		
Backwater length	km		
Coefficient of storage			
Regulation characteristics			
Maximum discharge at check flood level	m^3/s		
Corresponding downstream water level	m		
Maximum discharge at design flood level	m^3/s		
Corresponding downstream water level	m		
Minimum discharge	m^3/s		Minimum discharge means base load flow, minimum flow for navigation or minimum discharge for ecological and water use requirement
Corresponding downstream water level	m		

Table B (Continued)

Description	Unit	Quantity	Remarks
2. Flood control project			
Protected area (or towns, industrial and mining areas)	10^4hm^2 (or km^2)		
Design standard P or actual flood	%		Current standard ($P=$ %)
Design water level	m		
Check standard P or actual flood	%		
Check water level	m		
Safety discharge of river	m^3/s		
Design flood diversion discharge	m^3/s		Location of flood diversion gate
Design flood diversion water level	m		
Design water level for flood impoundment (detention)	m		Inside flood storage (detention) basin
Storage capacity for flood impoundment (detention)	10^8m^3		
3. Waterlogging control project			
Area	10^4hm^2 (or km^2)		
Design standard			Current standard ($P=$ %)
Drainage discharge	m^3/s		
Maximum water level in drainage receiving area	m		

Table B (Continued)

Description	Unit	Quantity	Remarks
Minimum water level in drainage receiving area	m		
Installed capacity of pump station	10^4 kW		
4. River channel and estuary training project			
Length of river reach to be trained	km		
Design flood standard P (or navigation grade)	%		Current standard ($P=$ %)
Design water level (or tide level)	m		
Check flood standard P	%		
Check water level (or tide level)	m		
Regulating water level	m		
Design discharge	m^3/s		
5. Irrigation project			
Design irrigation area (short-term)	$10^4 hm^2$		
Design irrigation area (long-term)	$10^4 hm^2$		
Design dependability of irrigation P	%		
Annual water diversion ($P=$ %)	$10^8 m^3$		Or annual average

Table B (Continued)

Description	Unit	Quantity	Remarks
Design diversion discharge	m^3/s		
Total installed capacity of pump station	$10^4 kW$		In the case of pumping irrigation
Total lift	m		
Annual power consumption for water pumping	$10^4 kW \cdot h$		
Design intake water level	m		Water intake from river channel
6. Water supply project			
Annual water diversion (short - term)	$10^8 m^3$		Annual average
Annual water diversion (long - term)	$10^8 m^3$		Annual average
Design diversion discharge	m^3/s		
Design dependability of water - supply P	%		
Annual diversion time	d		
Length of diversion line	km		
Total installed capacity of pump station	$10^4 kW$		
Total lift	m		
Annual power consumption for water pumping	$10^4 kW \cdot h$		
Design intake water level	m		Water intake from river channel

Table B (Continued)

Description	Unit	Quantity	Remarks
7. Hydropower project			
Installed capacity	10^4 kW		
Firm output	10^4 kW		
Average annual energy output	10^8 kW·h		
Annual utilization hours	h		
Regulation characteristics of reservoir			
Design intake water level	m		Conduit type hydropower station
Minimum intake water level	m		
Diversion discharge for power generation	m^3/s		
8. Reclamation project			
Area	$10^4 hm^2$ (or km^2)		
Design flood (tide) control standard P (or actual flood)	%		
Design flood (tide) control water level	m		
Design water-supply (irrigation) discharge	m^3/s		
Design drainage discharge	m^3/s		

Table B (Continued)

Description	Unit	Quantity	Remarks
Benefits			Including aquaculture, crop-planting, land development and so on
9. Navigation project			
Design passenger (freight) capacity	t/a		
Design maximum ship tonnage	t		
Maximum navigable water level upstream	m		
Minimum navigable water level upstream	m		
Maximum navigable water level downstream	m		
Minimum navigable water level downstream	m		
III. Inundation loss and land acquisition for permanent use			
1. Inundated land ($P=$ %)	10^4hm^2		
Where: cultivated land	10^4hm^2		List paddy field, dry farmland, grassland and so on respectively
2. Relocated population ($P=$ %)	person		
3. Houses in submerged area	m^2		
4. Forest in submerged area	hm^2		

Table B (Continued)

Description	Unit	Quantity	Remarks
5. Important specific facilities affected due to submerging			
6. Land acquisition for construction	hm²		
Where: cultivated land	hm²		
7. Land acquisition for management	hm²		
Where: cultivated land	hm²		
IV. Major structures and equipment			
1. Water retaining structures (dam, sluice, and dike)			
Type			
Foundation features			
Design value of ground motion parameter	g		
Basic seismic intensity			
Design seismic intensity			
Crest elevation (dam, sluice, and levee)	m		
Maximum dam (sluice and levee) height	m		
Crest length (dam, sluice, and levee)	m		

Table B (Continued)

Description	Unit	Quantity	Remarks
2. Water releasing structures (overflow weir, spillway, tunnel, bottom outlet, sluice, ⋯)			List each structure respectively
Type			
Foundation features			
Crest elevation of weir	m		
Length of overflow section (or dimension and number of flood discharge tunnels and sluice)	m		
Design flood discharge	m^3/s		
Check flood discharge	m^3/s		
3. Water diversion structure			
Design diversion discharge	m^3/s		
Maximum diversion discharge	m^3/s		
Invert elevation of intake	m		
Headrace: type			
Length	m		
Section dimensions	m		

Table B (Continued)

Description	Unit	Quantity	Remarks
Surge chamber (or forebay): type			
Main dimensions	m		
Penstock: type			
Number			
Length of each pipe	m		
Inner diameter	m		
4. Water conveyance works			
Design flow	m^3/s		
Water conveyance structures: type			
Length	m		
Section dimensions	m		
Crossing-structures: type			Aqueduct and inverted siphon
5. Powerhouse			
Type			
Main powerhouse dimensions ($L \times W \times H$)	$m \times m \times m$		
Setting elevation of turbine or pump	m		

Table B (Continued)

Description	Unit	Quantity	Remarks
6. Switchyard (converter station and substation)			
Type			
Area ($L \times W$)	m$\times$m		
7. Major electromechanical equipment			
Number of turbines (pumps)	set		
Type			
Rated output (input)	kW		
Number of generators (motors)	set		
Type			
Unit capacity	kW		
Number and type of main transformers			
8. Transmission line			
Voltage	kV		
Number of circuits			
Transmission distance	km		
9. Navigation structures			

Table B (Continued)

Description		Unit	Quantity	Remarks
Type				
Main dimensions		m		
Navigation discharge		m³/s		Maximum, minimum
10. Fish passing structures				
Type				
Main dimensions		m		
Discharge		m³/s		Maximum, minimum
11. Other structures (sandtrap, fish interception facilities, reservoir wharf, monitoring instrumentation, permanent houses, etc.)				
V. Construction				
1. Quantity of main works				
Open excavation	Earth	10⁴ m³		
	Rock	10⁴ m³		
Rock tunnel excavation		10⁴ m³		

Table B (Continued)

Description		Unit	Quantity	Remarks
Backfill	Earth	$10^4 \, m^3$		
	Rock	$10^4 \, m^3$		
Dry laid riprap		$10^4 \, m^3$		
Cemented riprap		$10^4 \, m^3$		
Concrete and reinforced concrete		$10^4 \, m^3$		
Hydraulic steel structures		t		
Curtain grouting		m		
Consolidation grouting		m		
2. Quantity of main construction materials				
Timber		m^3		
Cement		t		
Steel		t		Including rebar, anchor bar, and anchor rod.
3. Demand of labors				
Total man – days		10^4 man – day		

Table B (Continued)

Description	Unit	Quantity	Remarks
Peak number of labors	person		
4. Construction power and sources			
Power supply	kW		Indicate the power sources
5. Access (highway, railway, and diversion structure)			
Distance	km		
Freight volume	10^4 t		
6. River diversion for construction (mode, type, and scale)			
7. Construction duration			
Pre-construction duration	month		
Construction duration before initial operation	month		It means the date of starting water retaining, impounding, water supply, power generation of the first unit, and navigation
Total construction duration	month		
VI. Economic indices			

Table B (Continued)

Description	Unit	Quantity	Remarks
1. Engineering			
Civil works	10⁴ Yuan		
Electromechanical works	10⁴ Yuan		
Hydraulic steel works	10⁴ Yuan		
Temporary works	10⁴ Yuan		
Independent costs	10⁴ Yuan		
Total static investment	10⁴ Yuan		
Where: Physical contingency	10⁴ Yuan		
Price contingency	10⁴ Yuan		
2. Compensation for land acquisition and resettlement			
Total static investment	10⁴ Yuan		
Where: physical contingency	10⁴ Yuan		
Price contingency	10⁴ Yuan		
3. Environmental protection			

Table B (Continued)

Description	Unit	Quantity	Remarks
Total static investment	10^4 Yuan		
Where: physical contingency	10^4 Yuan		
Price contingency	10^4 Yuan		
4. Water and soil conservation			
Total static investment	10^4 Yuan		
Where: physical contingency	10^4 Yuan		
Price contingency	10^4 Yuan		
5. Total investment			
Total static investment	10^4 Yuan		
Where: physical contingency	10^4 Yuan		
Price contingency	10^4 Yuan		
Financing interest in construction duration	10^4 Yuan		
Total investment	10^4 Yuan		
Ⅶ. Comprehensive utilization economic indices			

Table B (Continued)

Description	Unit	Quantity	Remarks
Investment in reservoir storage capacity	Yuan/m^3		
Investment in river training	Yuan/km		
Investment in irrigation area	Yuan/hm^2		
Investment in reclamation area	Yuan/hm^2		
Water supply investment	Yuan/m^3		
Water supply cost	Yuan/m^3		
Investment in kilowatt	Yuan/kW		
Power generation cost	Yuan/(kW·h)		
Economic internal rate of return (EIRR)	%		
Financial internal rate of return (FIRR)	%		
Water price	Yuan/m^3		
Feed-in tariff	Yuan/(kW·h)		
Loan repayment period	a		
Other economic indices			

Explanation of Wording

Words in this standard	Equivalent expressions in specific situations	Strictness of requirement
Shall	It is necessary	Required (Words denoting a very strict or mandatory requirement)
Shall not	It is not allowed/permitted It is unacceptable	
Should	It is recommended It is advisable	Recommended (Words denoting a strict requirement under normal conditions)
Should not	It is not recommended It is not advisable	
May	It is suitable It is desirable It is preferable	Permitted (Words denoting a permission of a slight choice or an indication of the most suitable choice when conditions permit)
Need not	It is unnecessary It is not required	